SAT MATH ESSENTIALS

Published in the United States by LearningExpress, LLC, New York.

Library of Congress Cataloging-in-Publication Data:
Cernese, Richard.
SAT math essentials / Richard Cernese, Dave Smith.
p. cm.
Includes bibliographical references and index.
ISBN 1-57685-533-3 (alk. paper)
1. Mathematics—Examinations, questions, etc.
2. College entrance achievement tests—United States—Study guides.
3. Scholastic Assessment Test—Study guides. I. Smith, Dave.
II. Title.
QA43.C35 2006
510.76—dc22

2005027526

Printed in the United States of America

9 8 7 6 5 4 3 2 1

ISBN 1-57685-533-3

For more information or to place an order, contact LearningExpress at:
55 Broadway
8th Floor
New York, NY 10006

Or visit us at:
www.learnatest.com

Contents

INTRODUCTION		v
CHAPTER 1	Taking the SAT	1
CHAPTER 2	Preparing for SAT Math	7
CHAPTER 3	Math Pretest	13
CHAPTER 4	Techniques and Strategies	27
CHAPTER 5	Numbers and Operations Review	37
CHAPTER 6	Algebra Review	69
CHAPTER 7	Geometry Review	95
CHAPTER 8	Problem Solving	149
CHAPTER 9	Practice Test 1	173
CHAPTER 10	Practice Test 2	197
CHAPTER 11	Practice Test 3	221
GLOSSARY		245

Introduction

Whether you bought this book, borrowed it, received it as a gift, took it out of the library, stole it (not a good idea!), or are simply reading it in a book store, you're undoubtedly hoping to ace the Math sections of the SAT Reasoning Test. Well, you've come to the right place to get prepared! This book provides answers to any and all questions you may have about the Math sections of the SAT. To get the most benefit from the book, work through it from cover to cover. Every hour you put into preparing for the SAT will pay off on test day. Here is a breakdown of what to expect in each section of the book:

Chapter 1 is an introduction to the SAT. It answers basic questions you may have about the exam.

Chapter 2 provides information about what to expect on the Math sections of the SAT and how best to prepare for the SAT.

Chapter 3 is a math pretest. This test serves as a warm-up, giving you a flavor for the range of math questions found on the SAT. Answers and explanations follow the pretest.

Chapter 4 teaches strategies and techniques for acing the Math sections of the SAT.

Chapter 5 reviews concepts of numbers and operations and provides sample numbers and operations SAT questions with explanations.

Chapter 6 reviews algebra and provides sample algebra SAT questions with explanations.

Chapter 7 reviews geometry and provides sample geometry SAT questions with explanations.

Chapter 8 reviews problem-solving skills and provides sample SAT word problems with explanations.

Chapters 9, **10**, and **11** are Practice Tests 1, 2, and 3. These practice tests are similar to the Math sections of the SAT. Answers and explanations follow the practice tests.

The **Glossary** provides definitions of all key math terms used in this book.

CHAPTER

Taking the SAT

What Is the SAT?

The SAT Reasoning Test is a standardized test developed by the Educational Testing Service for The College Board, an association of colleges and schools. It contains questions that test skills in math, reading, and writing.

Why Take the SAT?

Most colleges require prospective students to submit SAT Reasoning Test scores as part of their applications. Colleges use SAT exam scores to help them evaluate the reading, writing, and math skills of prospective students. Therefore, it is important to do your best on the SAT so you can show colleges what you are capable of accomplishing.

▶ Who Takes the SAT?

The SAT Reasoning Test is the most common standardized test that high school students take when applying to college. In fact, approximately two million students take the SAT each year.

▶ Will My SAT Scores Determine Whether I Get into College?

No. Your SAT scores are only one small part of any college application. In other words, your SAT scores alone will not determine whether or not a college accepts you as part of its student body. Let's say that again, a little louder: YOUR SAT SCORES ALONE WILL NOT DETERMINE WHETHER OR NOT A COLLEGE ACCEPTS YOU AS PART OF ITS STUDENT BODY. Colleges look at individuals, not just test scores and grades. They want fascinating, curious, motivated people on their campuses, not a bunch of numbers.

When evaluating candidates, admissions officers look at your academic performance, but they also look at the rest of your life. What are your interests? How do you spend your time outside of school? What are your goals?

When you submit an application to college, you should make sure it shows what makes you a unique person. Colleges typically aim to fill their campuses with a diverse group of individuals. Think about what you can best offer to a college community. What are your strong points? Do you excel in music, theater, art, sports, academics, student government, community service, business, or other areas? It doesn't matter what your interests are. It only matters that you have them. Let your best qualities shine through in your application and you can be confident that you are presenting yourself as a strong possible candidate for admission.

So, don't sweat the SAT. Getting nervous about it won't help you anyway. As long as you follow through with your plan to prepare for it, your score can help you become an attractive candidate.

▶ When Do I Take the SAT?

The SAT is offered on Saturday mornings several times a year. Your high school guidance office can give you a schedule. You can also find a schedule online at www.collegeboard.com. Please note that Sunday administrations will occur the day after each Saturday test date for students who cannot test on Saturday for religious reasons.

▶ How Many Times Should I Take the SAT?

The number of times you take the SAT is up to you. You may register and take the exam as often as you wish. Most colleges will not hold an initial lower score against you, and some will be impressed by a substantially improved score, so taking the SAT twice or three times with the goal of raising your score is recommended if you think you can do better. However, some students prepare hard for their first SAT and feel satisfied with their initial score.

Regardless, you shouldn't take the SAT more than three times. Chances are your score will not change significantly on your fourth test. If you are still disappointed after your third score, your time, money, and energy will be better spent on other parts of your college application.

But no matter how many times you have taken the SAT, you're smart to be using this book. The only way to raise your SAT score is through preparation and practice.

▶ Where Is the SAT Given?

Many high school and college campuses host the SATs. When you register, you will be given a list of sites in your local area, and you can pick one that is comfortable and convenient for you.

▶ Where Do I Sign Up for the SAT?

To sign up for the SAT, you can:

1. Register online at The College Board's website: www.collegeboard.com.
2. Get the SAT Registration Bulletin from your high school guidance office. The Bulletin contains a registration form and other important information about the exam. If you are retaking the exam, you can also register by phone at 800-SAT-SCORE.

▶ How Long Is the SAT?

The SAT takes three hours and 45 minutes. In addition to the testing time, you will get two or three five- to ten-minute breaks between sections of the exam. You will also spend up to an additional hour filling out forms. Overall, you can expect to be at the testing location for about four and a half hours.

▶ What Is Tested on the SAT?

The SAT has approximately 160 questions divided into eight test sections:

- three critical reading sections
  - two 25-minute sections
  - one 20-minute section
- three math sections
  - two 25-minute sections
  - one 20-minute section
- two writing sections
  - one 35-minute multiple-choice section
  - one 25-minute essay

Your scores on these eight sections make up your SAT scores.

In addition to the core eight sections, there is one unscored "variable," or "equating," section that the test writers use to evaluate new questions before including them on future SATs. Thus, you will actually complete a total of nine sections on test day. But it will be impossible for you to tell which section is the variable section: It can be critical reading, multiple-choice writing, or math, and it can appear in any place on the exam. So although the variable section does not affect your SAT score, you must treat each section as if it counts.

▶ In What Order Are the Sections Tested?

The writing essay is always the first section of the SAT. The multiple-choice writing section is always the last section. The remaining sections can appear in any order.

▶ How Is the SAT Scored?

SAT scores range from 600–2400. You can score a minimum of 200 and a maximum of 800 on each subject: math, critical reading, and writing.

A computer scores the math questions. For the multiple-choice math questions, the computer counts the number of correct answers and gives one point for each. Then it counts your incorrect answers and deducts one-quarter point from the total of your correct answers. For the grid-in math questions, the computer counts the number of correct answers and gives one point for each. No points are subtracted for incorrect answers to the grid-in questions. If the score that results from the subtraction is a fraction of a point,

Four Steps to Scoring Math Questions on the SAT

For multiple-choice questions:

1. Correct answers are added: 1 point for each correct answer.
2. Incorrect answers are subtracted: $\frac{1}{4}$ point for each wrong answer.
3. Your raw score is the result of adding correct answers, subtracting incorrect answers, and then rounding the result to the nearest whole number.

For grid-in questions:

1. Right answers are added: 1 point for each correct answer.
2. Wrong answers receive zero points: No points are subtracted.
3. Your raw score is the total number of correct answers.

Once questions are scored, raw scores are converted to scaled scores, using an equating process.

your score is rounded to the nearest whole number. Your raw score for the math sections is then converted to a scaled score (between 200 and 800), using the statistical process of equating.

▶ Math Score Reporting

The College Board will send you a report on your scores. They will also send your scores to any schools (up to four) you requested on your application. Colleges, naturally, are used to seeing these reports, but they can be confusing to everybody else. Here's how you look at them:

You will see your scaled math score in a column headed *Score*. There are also columns titled *Score Range* and *Percentiles College-bound Seniors*. The information in these columns can be useful in your preparations for college.

Score Range

Immediately following your total scaled math score, there is a score range, which is a 60-point spread. Your actual scaled score falls right in the middle of this range.

Based on experience, The College Board believes that if you retake the SAT without further preparation, you are unlikely to move up or down more than thirty points within each subject tested. In other words, if you scored a 550 in math on your first SAT, chances are you won't score less than 520 or more than 580 in math if you take the exam again without any extra preparation. For this reason, it presents your score within a 60-point range to suggest that those are the range of scores that you could expect to get on the SAT.

Keep in mind that The College Board believes your score won't change if you retake the SAT without further preparation. *With further preparation*, such as using this book, your score can improve by much more than 30 points.

Percentile

Your score report will also include two percentile rankings. The first measures your SAT scores against those of all students nationwide who took the test. The second measures your scores against only the students in your state who took the test.

The higher your percentile ranking the better. For example, if you receive a 65 in the national category and a 67 in the state category, your scores were better than 65% of students nationwide and better than 67% in your state. In other words, of every 100 students who took the test in your state, you scored higher than 67 of them.

Additional Score Information

Along with information about your scaled score, The College Board also includes information about your raw score. The raw score tells you how well you did on each type of critical reading, math, and writing question—how many questions you answered correctly, how many you answered incorrectly, and how many you left blank. You can use this information to determine whether you can improve on a particular type of question. If you have already taken the SAT, use this information to see where you need to focus your preparation.

You will also receive information about the colleges or universities to which you have asked The College Board to report your scores. This information will include typical SAT scores of students at these schools as well as other admission policies and financial information.

When you look at SAT scores for a particular school, keep in mind that those scores are not the only criterion for admission to or success at any school. They are only part of any application package. Also, your SAT report includes only the score range for the middle 50% of first-year students at each school. It tells you that 25% of the first-year students scored higher than that range and the 25% scored below that range. So if your score falls below that range for a particular school, don't think admissions officers automatically won't be interested in you. In fact, one-fourth of their first-year students scored below that range.

CHAPTER

Preparing for SAT Math

▶ What to Expect

There are three Math sections on the SAT: two 25-minute sections and one 20-minute section. The Math sections contain two types of questions: five-choice and grid-ins.

▶ Five-Choice Questions

The **five-choice** questions, which are multiple-choice questions, present a question followed by five answer choices. You choose which answer choice you think is the best answer to the question. Questions test the following subject areas: numbers and operations (i.e., arithmetic), geometry, algebra and functions, statistics and data analysis, and probability. About 90% of the questions on the Math section are five-choice questions.

Here is an example:

1. By how much does the product of 13 and 20 exceed the product of 25 and 10?

a. 1
b. 5
c. 10
d. 15
e. 20

1. ⓐ ⓑ ● ⓓ ⓔ

Five-choice questions test your mathematical reasoning skills. They require you to apply various math techniques for each problem.

▶ Grid-In Questions

Grid-in questions are also called *student-produced responses.* There are approximately ten grid-in questions on the entire exam. Grid-in questions do not provide you with answer choices. Instead, a grid-in question asks you to solve a math problem and then enter the correct answer on your answer sheet by filling in numbered ovals on a grid.

You can fill in whole numbers, fractions, and decimals on the grids. Examples follow.

Whole Numbers

If your answer is 257, fill in the number ovals marked 2, 5, and 7:

2	5	7	
	/	/	
.	.	.	.
	0	0	0
1	1	1	1
●	2	2	2
3	3	3	3
4	4	4	4
5	●	5	5
6	6	6	6
7	7	●	7
8	8	8	8
9	9	9	9

Fractions

If your answer is $\frac{4}{9}$, fill in the number ovals marked 4 and 9 and a fraction symbol (/) in between.

	4	/	9
	/	●	
.	.	.	.
	0	0	0
1	1	1	1
2	2	2	2
3	3	3	3
4	●	4	4
5	5	5	5
6	6	6	6
7	7	7	7
8	8	8	8
9	9	9	●

Note that all mixed numbers should be written as improper fractions. For example, $5\frac{3}{5}$ should be filled in as 28/5.

Decimals

If your answer is 3.06, fill in the number ovals marked 3, 0, and 6 with a decimal point in between the 3 and the 0.

3	.	0	6
	/	/	
.	●	.	.
	0	●	0
1	1	1	1
2	2	2	2
●	3	3	3
4	4	4	4
5	5	5	5
6	6	6	●
7	7	7	7
8	8	8	8
9	9	9	9

Using the Right Columns

The scoring machine gives you credit for your answer no matter which columns you use. For example, all three of these grids would be scored correct for the answer 42:

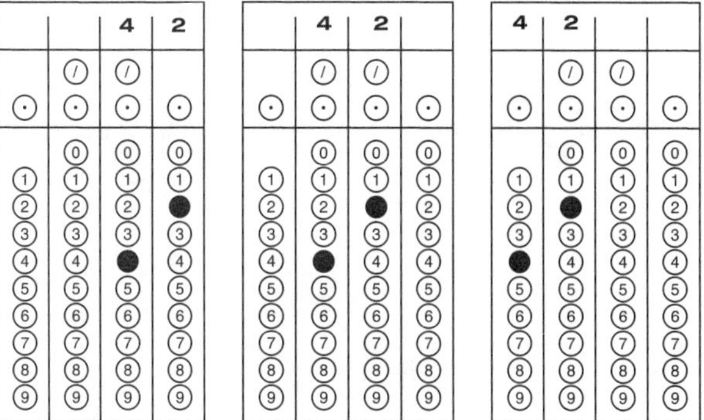

However, so that you don't confuse yourself, we recommend using the placement on the left. And be sure to leave unused grid columns blank.

Units

Grids do not have ovals for units, such as $ or °, so do not write them in. If you need to write an answer that includes units, simply leave the units out. For example, you would fill in $4.97 as 4.97 and 90° as 90.

Percents

If you determine an answer as a percent, such as 50%, do not fill in 50% on the grid. The grid does not have a percent symbol (%). Instead, convert all percents to fractions or decimals before filling in the grid. For example, 50% should be filled in as .50 or 1/2.

Ratios

The grid also does not have a ratio symbol (:). For grid-in items, write all ratios as fractions or decimals. For example, 1:4 or "1 to 4" should be filled in as 1/4 or .25.

Negative Numbers and Variables

You cannot mark a negative number or a variable on a grid. Therefore, if you solve a grid-in problem and determine an answer that includes a variable or a negative sign, you know your answer must be wrong! Solve it again!

Fill Those Ovals!

As you can see in the samples, there is space to write your answer in number form at the top of each grid above the ovals. However, grid-in questions are scored by machine, and the machines only read the ovals. SO YOU MUST FILL IN THE OVALS IN ORDER TO GET CREDIT! You actually don't even need to hand-write the answer at the top. But it's usually a good idea to write your answer before filling in the ovals so that you don't make an error.

Become Familiar with Grids!

Be sure you are very familiar with how to fill in a grid before you take the real SAT. You don't want to waste any test time trying to figure out how to fill in the grids.

▶ How to Prepare

Use the following strategies to maximize the effectiveness of your SAT preparation.

Take the Time

The more time you can spend preparing for the SAT, the better prepared you will be. However, you don't need to spend several hours at once to study well. Between now and test day, dedicate one or two hours a day to using this book. You'll be surprised at how much you can accomplish. Spending an hour a day over a few months will be much more beneficial than spending five hours a day during the week before the exam.

Don't Cram

Just as you don't train to run a marathon by waiting until the last minute and then running twenty miles a day for five days before the race, you cannot prepare most effectively for the SAT by waiting until the last

SAT Math at a Glance

Math Sections

- two 25-minute sections
- one 20-minute math section
- total of 70 minutes for math sections

Math Questions

- 90% are multiple-choice questions; you must choose an answer from five answer choices
- about ten questions are grid-in questions; you must determine the answer without answer choices

Math Concepts Tested

- numbers and operations (i.e., arithmetic)
- geometry
- algebra and functions
- statistics and data analysis
- probability

minute to study. Your brain works best when you give it a relatively small chunk of information, let it rest and process, and then give it another small chunk.

Stay Focused

During your study time, keep the TV and various computer programs (such as AIM) off, don't answer the phone, and stay focused on your work. Don't give yourself the opportunity to be distracted.

Find the Right Time and Place

Some times of the day may be better times for you to study than others. Some places may be more conducive to good studying than others. Choose a time to study when you are alert and can concentrate easily. Choose a place to study where you can be comfortable and where there aren't any distractions. Ideally, you should choose the perfect time and place and use them every day. Get into a routine, and you'll find that studying for the SAT will be no different than taking a shower or eating dinner.

Because the SAT is given early on Saturday mornings, you may want to spend some of your study time early in the morning—especially in the weeks leading up to the test—so you can accustom yourself to thinking about SAT questions at that time of day. Even better would be to dedicate several of the Saturday mornings before the test to SAT preparation. Get yourself used to walking up early on Saturdays and working on the SAT. Then, when test day arrives, getting up early and concentrating on SAT questions will seem like no big deal.

Reward Yourself

Studying is hard work. That's why studying is so beneficial. One way you can help yourself stay motivated to study is to set up a system of rewards. For example, if you keep your commitment to study for an hour in the afternoon, reward yourself afterward, perhaps with a glass of lemonade or the time to read a magazine. If you stay on track all week, reward yourself with a movie with friends or something else you enjoy. The point is to keep yourself dedicated to your work without letting the SAT become all you think about. Remember: If you put in the hard work, you'll enjoy your relaxation time even more.

Use Additional Study Sources

This book will give you a solid foundation of knowledge about the math sections of the SAT. However, you might also benefit from other LearningExpress books such as *Practical Math Success in 20 Minutes a Day* and *1001 Math Questions.*

Take Real Practice Tests

It is essential that you obtain the book *10 Real SATs,* published by The College Board. This book is the only source for actual retired SATs. Make sure you take at least one real retired SAT before test day. The more familiar you can become with the look and feel of a real SAT, the fewer surprises there will be on test day.

Memorize the Directions

The directions found on SATs are the same from test to test, so memorize the directions on the practice tests in the *10 Real SATs* book so you won't have to read the directions on test day. This will save you a lot of time. While some students will be reading through the directions, you can be working on the first question.

▶ How to Use This Book

You will need the following materials while working with this book:

- a notebook or legal pad dedicated to your SAT work
- pencils (and a pencil sharpener) or pens
- a four-function, scientific, or graphing calculator (Note: Calculators are not required for the SAT, but they are recommended, so you should practice using one when answering the questions in this book.)
- different-colored highlighters for highlighting important ideas
- paper clips or sticky note pads for marking pages you want to return to
- a calendar

you may, of course, use this book however you like. Perhaps you need only to study one area of math or want only to take the practice tests. However, for the best results from this book, follow this guide:

1. Take the pretest in Chapter 3. This is a short test with questions similar to those you will see on the SAT. This pretest will give you a flavor of the types of math questions the SAT includes. Don't worry if any of the questions confuse you. They are designed only to get your feet wet before you work through the rest of the book.
2. Work through Chapters 4–8. These chapters are the meat of the book and will give you techniques and strategies for answering SAT math questions successfully. They will also review the math skills and concepts you need to know for the SAT.
3. Take the practice tests in Chapters 9, 10, and 11. Make sure to read through the answers and explanations when you finish. Review your errors to determine if you need to study any parts of the book again.

CHAPTER

3 Math Pretest

The pretest contains questions similar to those found on the SAT. Take the pretest to familiarize yourself with the types of questions you will be preparing yourself for as you study this book.

Do not time yourself on the pretest. Solve each question as best you can. When you are finished with the test, review the answers and explanations that immediately follow the test. Make note of the kinds of errors you made and focus on these problems while studying the rest of this book.

1. ⓐ ⓑ ⓒ ⓓ ⓔ	6. ⓐ ⓑ ⓒ ⓓ ⓔ	11. ⓐ ⓑ ⓒ ⓓ ⓔ
2. ⓐ ⓑ ⓒ ⓓ ⓔ	7. ⓐ ⓑ ⓒ ⓓ ⓔ	12. ⓐ ⓑ ⓒ ⓓ ⓔ
3. ⓐ ⓑ ⓒ ⓓ ⓔ	8. ⓐ ⓑ ⓒ ⓓ ⓔ	13. ⓐ ⓑ ⓒ ⓓ ⓔ
4. ⓐ ⓑ ⓒ ⓓ ⓔ	9. ⓐ ⓑ ⓒ ⓓ ⓔ	14. ⓐ ⓑ ⓒ ⓓ ⓔ
5. ⓐ ⓑ ⓒ ⓓ ⓔ	10. ⓐ ⓑ ⓒ ⓓ ⓔ	15. ⓐ ⓑ ⓒ ⓓ ⓔ

16. **17.** **18.** **19.** **20.**

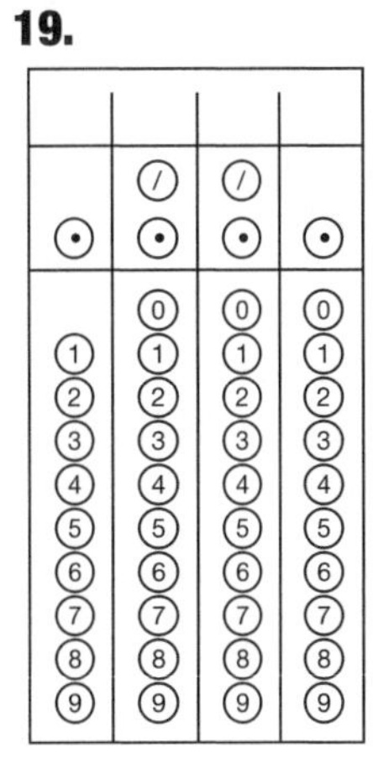

21. **22.** **23.** **24.** **25.**

26. **27.** **28.** **29.** **30.**

1. If $w = \frac{1}{8}$, then $w^{\frac{2}{3}} = ?$

a. $\frac{1}{2}$

b. $\frac{1}{4}$

c. $\frac{1}{8}$

d. $\frac{1}{12}$

e. $\frac{1}{64}$

2. Ben is three times as old as Samantha, who is two years older than half of Michele's age. If Michele is 12, how old is Ben?

a. 8

b. 18

c. 20

d. 24

e. 36

3. The expression $x^2 - 8x + 12$ is equal to 0 when $x = 2$ and when $x = ?$

a. –12

b. –6

c. –2

d. 4

e. 6

4. Mia ran 0.60 km on Saturday, 0.75 km on Sunday, and 1.4 km on Monday. How many km did she run in total?

a. $1\frac{1}{5}$ km

b. $1\frac{3}{4}$ km

c. $2\frac{1}{4}$ km

d. $2\frac{3}{4}$ km

e. $3\frac{1}{2}$ km

5.

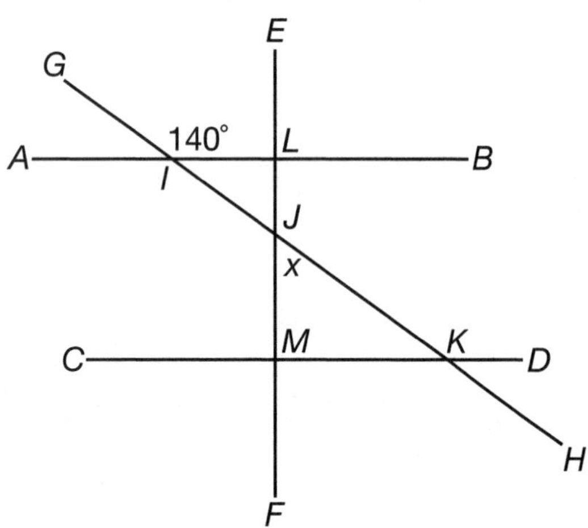

In the diagram above, line *AB* is parallel to line *CD*, and line *EF* is perpendicular to line *CD*. What is the measure of angle *x*?

a. 40 degrees
b. 45 degrees
c. 50 degrees
d. 60 degrees
e. 80 degrees

6. The area of circle *A* is 6.25π in^2. If the radius of the circle is doubled, what is the new area of circle *A*?

a. 5π in^2
b. 12.5π in^2
c. 25π in^2
d. 39.0625π in^2
e. 156.25π in^2

7. David draws a line that is 13 units long. If (–4,1) is one endpoint of the line, which of the following could be the other endpoint?

a. (1,13)
b. (9,14)
c. (3,7)
d. (5,12)
e. (13,13)

8. The expression $(\frac{a^2}{b^3})(\frac{a^{-2}}{b^{-3}}) = ?$

a. 0

b. 1

c. $(\frac{a^{-4}}{b^{-9}})$

d. $(\frac{b^9}{a^4})$

e. b^{-9}

9.

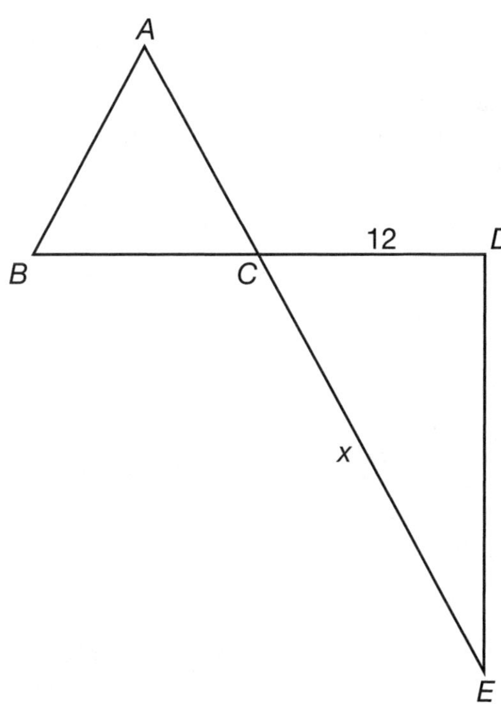

If triangle ABC in the figure above is an equilateral triangle and D is a right angle, find the value of x.

a. $6\sqrt{3}$
b. $8\sqrt{3}$
c. $12\sqrt{2}$
d. 13
e. 24

10. If 10% of x is equal to 25% of y, and $y = 16$, what is the value of x?

a. 4
b. 6.4
c. 24
d. 40
e. 64

11.

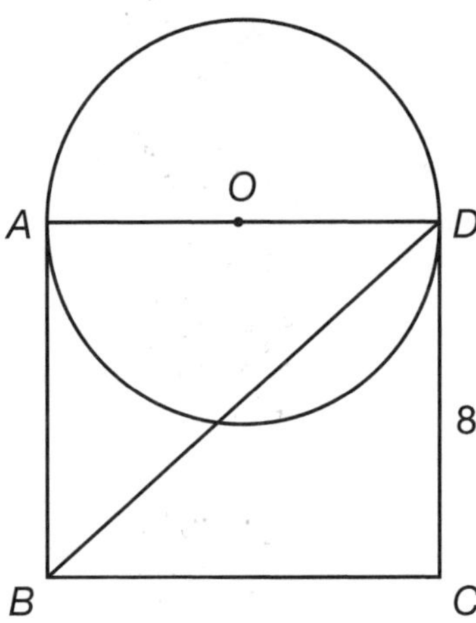

Triangle *BDC*, shown above, has an area of 48 square units. If *ABCD* is a rectangle, what is the area of the circle in square units?

a. 6π square units
b. 12π square units
c. 24π square units
d. 30π square units
e. 36π square units

12. If the diagonal of a square measures $16\sqrt{2}$ cm, what is the area of the square?

a. $32\sqrt{2}\ \text{cm}^2$
b. $64\sqrt{2}\ \text{cm}^2$
c. $128\ \text{cm}^2$
d. $256\ \text{cm}^2$
e. $512\ \text{cm}^2$

13. If $m > n$, which of the following must be true?

a. $\frac{m}{2} > \frac{n}{2}$
b. $m^2 > n^2$
c. $mn > 0$
d. $|m| > |n|$
e. $mn > -mn$

14. Every 3 minutes, 4 liters of water are poured into a 2,000-liter tank. After 2 hours, what percent of the tank is full?

a. 0.4%
b. 4%
c. 8%
d. 12%
e. 16%

15. What is the perimeter of the shaded area, if the shape is a quarter circle with a radius of 8?

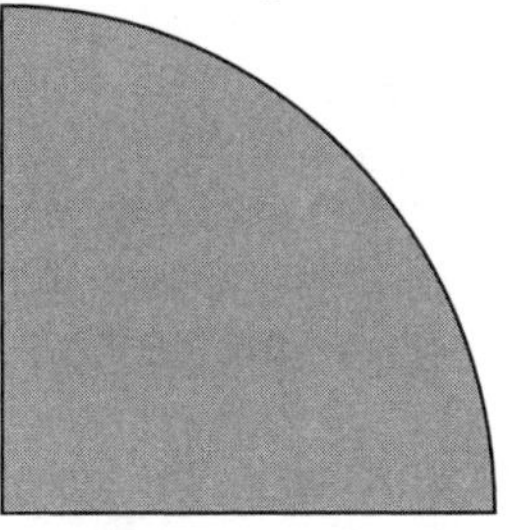

a. 2π
b. 4π
c. $2\pi + 16$
d. $4\pi + 16$
e. 16π

16. Melanie compares two restaurant menus. The Scarlet Inn has two appetizers, five entrées, and four desserts. The Montgomery Garden offers three appetizers, four entrées, and three desserts. If a meal consists of an appetizer, an entrée, and a dessert, how many more meal combinations does the Scarlet Inn offer?

17.

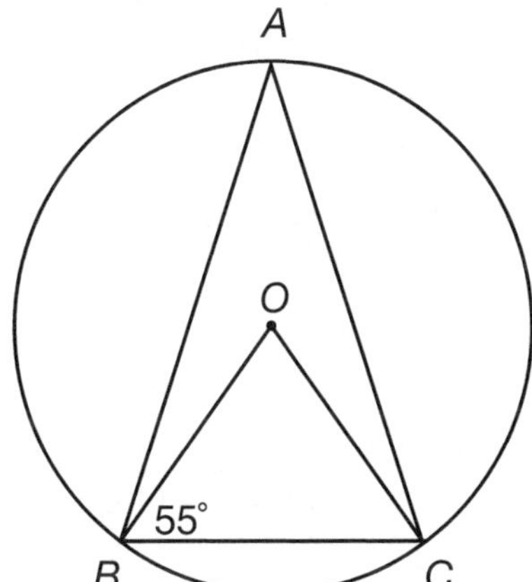

In the diagram above, angle *OBC* is congruent to angle *OCB*. How many degrees does angle *A* measure?

18. Find the positive value that makes the function $f(a) = \frac{4a^2 + 12a + 9}{a^2 - 16}$ undefined.

19. Kiki is climbing a mountain. His elevation at the start of today is 900 feet. After 12 hours, Kiki is at an elevation of 1,452 feet. On average, how many feet did Kiki climb per hour today?

20. Freddie walks three dogs, which weigh an average of 75 pounds each. After Freddie begins to walk a fourth dog, the average weight of the dogs drops to 70 pounds. What is the weight in pounds of the fourth dog?

21. Kerry began lifting weights in January. After 6 months, he can lift 312 pounds, a 20% increase in the weight he could lift when he began. How much weight could Kerry lift in January?

22.

RECYCLER	ALUMINUM	CARDBOARD	GLASS	PLASTIC
x	.06/pound	.03/pound	.08/pound	.02/pound
y	.07/pound	.04/pound	.07/pound	.03/pound

If you take recyclables to whichever recycler will pay the most, what is the greatest amount of money you could get for 2,200 pounds of aluminum, 1,400 pounds of cardboard, 3,100 pounds of glass, and 900 pounds of plastic?

23. The sum of three consecutive integers is 60. Find the least of these integers.

24. What is the sixth term of the sequence: $\frac{1}{3}, \frac{1}{2}, \frac{3}{4}, \frac{9}{8}, \ldots$?

25. The graph of the equation $\frac{2x-3}{3y} = 4$ crosses the y-axis at the point $(0,a)$. Find the value of a.

26. The angles of a triangle are in the ratio 1:3:5. What is the measure, in degrees, of the largest angle of the triangle?

27. Each face of a cube is identical to two faces of rectangular prism whose edges are all integers larger than one unit in measure. If the surface area of one face of the prism is 9 square units and the surface area of another face of the prism is 21 square units, find the possible surface area of the cube.

28. The numbers 1 through 40 are written on 40 cards, one number on each card, and stacked in a deck. The cards numbered 2, 8, 12, 16, 24, 30, and 38 are removed from the deck. If Jodi now selects a card at random from the deck, what is the probability that the card's number is a multiple of 4 and a factor of 40?

29. Suppose the amount of radiation that could be received from a microwave oven varies inversely as the square of the distance from it. How many feet away must you stand to reduce your potential radiation exposure to $\frac{1}{16}$ the amount you could have received standing 1 foot away?

30. The variable x represents Cindy's favorite number and the variable y represents Wendy's favorite number. For this given x and y, if $x > y > 1$, x and y are both prime numbers, and x and y are both whole numbers, how many whole number factors exist for the product of the girls' favorite numbers?

▶ Answers

1. b. Substitute $\frac{1}{8}$ for w. To raise $\frac{1}{8}$ to the exponent $\frac{2}{3}$, square $\frac{1}{8}$ and then take the cube root. $\frac{1}{8}^2 = \frac{1}{64}$, and the cube root of $\frac{1}{64} = \frac{1}{4}$.

2. d. Samantha is two years older than half of Michele's age. Since Michele is 12, Samantha is $(12 \div 2) + 2 = 8$. Ben is three times as old as Samantha, so Ben is 24.

3. e. Factor the expression $x^2 - 8x + 12$ and set each factor equal to 0:
$x^2 - 8x + 12 = (x - 2)(x - 6)$
$x - 2 = 0$, so $x = 2$
$x - 6 = 0$, so $x = 6$

4. d. Add up the individual distances to get the total amount that Mia ran; $0.60 + 0.75 + 1.4 = 2.75$ km. Convert this into a fraction by adding the whole number, 2, to the fraction $\frac{75}{100} \div \frac{25}{25} = \frac{3}{4}$. The answer is $2\frac{3}{4}$ km.

5. c. Since lines *EF* and *CD* are perpendicular, triangles *ILJ* and *JMK* are right triangles. Angles *GIL* and *JKD* are alternating angles, since lines *AB* and *CD* are parallel and cut by transversal *GH*. Therefore, angles *GIL* and *JKD* are congruent—they both measure 140 degrees. Angles *JKD* and *JKM* form a line. A line has 180 degrees, so the measure of angle $JKM = 180 - 140 = 40$ degrees. There are also 180 degrees in a triangle. Right angle *JMK*, 90 degrees, angle *JKM*, 40 degrees, and angle x form a triangle. Angle x is equal to $180 - (90 + 40) = 180 - 130 = 50$ degrees.

6. c. The area of a circle is equal to πr^2, where r is the radius of the circle. If the radius, r, is doubled $(2r)$, the area of the circle increases by a factor of four, from πr^2 to $\pi(2r)^2 = 4\pi r^2$. Multiply the area of the old circle by four to find the new area of the circle:
$6.25\pi \text{ in}^2 \times 4 = 25\pi \text{ in}^2$.

7. a. The distance formula is equal to $\sqrt{((x_2 - x_1)^2 + (y_2 - y_1)^2)}$. Substituting the endpoints (–4,1) and (1,13), we find that $\sqrt{((-4 - 1)^2 + (1 - 13)^2)} = \sqrt{((-5)^2 + (-12)^2)} = \sqrt{25 + 144} = \sqrt{169} = 13$, the length of David's line.

8. b. A term with a negative exponent in the numerator of a fraction can be rewritten with a positive exponent in the denominator, and a term with a negative exponent in the denominator of a fraction can be rewritten with a positive exponent in the numerator. $(\frac{a^{-2}}{b^{-3}}) = (\frac{b^3}{a^2})$. When $(\frac{a^2}{b^3})$ is multiplied by $(\frac{b^3}{a^2})$, the numerators and denominators cancel each other out and you are left with the fraction $\frac{1}{1}$, or 1.

9. e. Since triangle *ABC* is equilateral, every angle in the triangle measures 60 degrees. Angles *ACB* and *DCE* are vertical angles. Vertical angles are congruent, so angle *DCE* also measures 60 degrees. Angle *D* is a right angle, so *CDE* is a right triangle. Given the measure of a side adjacent to angle *DCE*, use the cosine of 60 degrees to find the length of side *CE*. The cosine is equal to $\frac{\text{(adjacent side)}}{\text{(hypotenuse)}}$, and the cosine of 60 degrees is equal to $\frac{1}{2}$; $\frac{12}{x} = \frac{1}{2}$, so $x = 24$.

10. d. First, find 25% of y; $16 \times 0.25 = 4$. 10% of x is equal to 4. Therefore, $0.1x = 4$. Divide both sides by 0.1 to find that $x = 40$.

11. e. The area of a triangle is equal to $(\frac{1}{2})bh$, where b is the base of the triangle and h is the height of the triangle. The area of triangle *BDC* is 48 square units and its height is 8 units.
$48 = \frac{1}{2}b(8)$
$48 = 4b$
$b = 12$
The base of the triangle, *BC*, is 12. Side *BC* is equal to side *AD*, the diameter of the circle.

The radius of the circle is equal to 6, half its diameter. The area of a circle is equal to πr^2, so the area of the circle is equal to 36π square units.

12. d. The sides of a square and the diagonal of a square form an isosceles right triangle. The length of the diagonal is $\sqrt{2}$ times the length of a side. The diagonal of the square is $16\sqrt{2}$ cm, therefore, one side of the square measures 16 cm. The area of a square is equal to the length of one side squared: $(16 \text{ cm})^2 = 256 \text{ cm}^2$.

13. a. If both sides of the inequality $\frac{m}{2} > \frac{n}{2}$ are multiplied by 2, the result is the original inequality, $m > n$. m^2 is not greater than n^2 when m is a positive number such as 1 and n is a negative number such as –2. mn is not greater than zero when m is positive and n is negative. The absolute value of m is not greater than the absolute value of n when m is 1 and n is –2. The product mn is not greater than the product $-mn$ when m is positive and n is negative.

14. c. There are 60 minutes in an hour and 120 minutes in two hours. If 4 liters are poured every 3 minutes, then 4 liters are poured 40 times ($120 \div 3$); $40 \times 4 = 160$. The tank, which holds 2,000 liters of water, is filled with 160 liters; $\frac{160}{2{,}000} = \frac{8}{100}$. 8% of the tank is full.

15. d. The curved portion of the shape is $\frac{1}{4}\pi d$, which is 4π. The linear portions are both the radius, so the solution is simply $4\pi + 16$.

16. 4 Multiply the number of appetizers, entrées, and desserts offered at each restaurant. The Scarlet Inn offers $(2)(5)(4) = 40$ meal combinations, and the Montgomery Garden offers $(3)(4)(3) = 36$ meal combinations. The Scarlet Inn offers four more meal combinations.

17. 35 Angles OBC and OCB are congruent, so both are equal to 55 degrees. The third angle in the triangle, angle O, is equal to $180 - (55 + 55) = 180 - 110 = 70$ degrees. Angle O is a central angle; therefore, arc BC is also equal to 70 degrees. Angle A is an inscribed angle. The measure of an inscribed angle is equal to half the measure of its intercepted arc. The measure of angle $A = 70 \div 2 = 35$ degrees.

18. 4 The function $f(a) = \frac{(4a^2 + 12a + 9)}{(a^2 - 16)}$ is undefined when its denominator is equal to zero; $a^2 - 16$ is equal to zero when $a = 4$ and when $a = -4$. The only positive value for which the function is undefined is 4.

19. 46 Over 12 hours, Kiki climbs $(1{,}452 - 900) = 552$ feet. On average, Kiki climbs $(552 \div 12) = 46$ feet per hour.

20. 55 The total weight of the first three dogs is equal to $75 \times 3 = 225$ pounds. The weight of the fourth dog, d, plus 225, divided by 4, is equal to the average weight of the four dogs, 70 pounds:

$\frac{d + 225}{4} = 70$

$d + 225 = 280$

$d = 55$ pounds

21. 260 The weight Kerry can lift now, 312 pounds, is 20% more, or 1.2 times more, than the weight, w, he could lift in January:

$1.2w = 312$

$w = 260$ pounds

22. 485 2,200(0.07) equals \$154; 1,400(0.04) equals \$56; 3,100(0.08) equals \$248; 900(0.03) equals \$27. Therefore, \$154 + \$56 + \$248 + \$27 = \$485.

23. 19 Let x, $x + 1$, and $x + 2$ represent the consecutive integers. The sum of these integers is 60: $x + x + 1 + x + 2 = 60$, $3x + 3 = 60$, $3x = 57$, $x = 19$. The integers are 19, 20, and 21, the smallest of which is 19.

24. $\frac{81}{32}$ Each term is equal to the previous term multiplied by $\frac{3}{2}$. The fifth term in the sequence is $\frac{9}{8} \times \frac{3}{2} = \frac{27}{16}$, and the sixth term is $\frac{27}{16} \times \frac{3}{2} = \frac{81}{32}$.

25. $-\frac{1}{4}$ The question is asking you to find the y-intercept of the equation $\frac{2x-3}{3y} = 4$. Multiply both sides by 3y and divide by 12: $y = \frac{1}{6}x - \frac{1}{4}$. The graph of the equation crosses the y-axis at $(0,-\frac{1}{4})$.

26. 100 Set the measures of the angles equal to $1x$, $3x$, and $5x$. The sum of the angle measures of a triangle is equal to 180 degrees:

$1x + 3x + 5x = 180$

$9x = 180$

$x = 20$

The angles of the triangle measure 20 degrees, 60 degrees, and 100 degrees.

27. 54 One face of the prism has a surface area of nine square units and another face has a surface area of 21 square units. These faces share a common edge. Three is the only factor common to 9 and 21 (other than one), which means that one face measures three units by three units and the other measures three units by seven units. The face of the prism that is identical to the face of the cube is in the shape of a square, since every face of a cube is in the shape of a square. The surface area of the square face is equal to nine square units, so surface area of one face of the cube is nine square units. A cube has six faces, so the surface area of the cube is $9 \times 6 = 54$ square units.

28. $\frac{1}{11}$ Seven cards are removed from the deck of 40, leaving 33 cards. There are three cards remaining that are both a multiple of 4 and a factor of 40: 4, 20, and 40. The probability of selecting one of those cards is $\frac{3}{33}$ or $\frac{1}{11}$.

29. 4 We are seeking D = number of feet away from the microwave where the amount of radiation is $\frac{1}{16}$ the initial amount. We are given: radiation varies inversely as the square of the distance or: $R = 1 \div D^2$. When $D = 1$, $R = 1$, so we are looking for D when $R = \frac{1}{16}$. Substituting: $\frac{1}{16} = 1 \div D^2$. Cross multiplying: $(1)(D^2) = (1)(16)$. Simplifying: $D^2 = 16$, or $D = 4$ feet.

30. 4 The factors of a number that is whole and prime are 1 and itself. For this we are given x and y, $x > y > 1$ and x and y are both prime. Therefore, the factors of x are 1 and x, and the factors of y are 1 and y. The factors of the product xy are 1, x, y, and xy. For a given x and y under these conditions, there are four factors for xy, the product of the girls' favorite numbers.

CHAPTER

Techniques and Strategies

The next four chapters will help you review all the mathematics you need to know for the SAT. However, before you jump ahead, make sure you first read and understand this chapter thoroughly. It includes techniques and strategies that you can apply to all SAT math questions.

▶ All Tests Are Not Alike

The SAT is not like the tests you are used to taking in school. It may test the same skills and concepts that your teachers have tested you on, but it tests them in different ways. Therefore, you need to know how to approach the questions on the SAT so that they don't surprise you with their tricks.

▶ The Truth about Multiple-Choice Questions

Many students think multiple-choice questions are easier than other types of questions because, unlike other types of questions, they provide you with the correct answer. You just need to figure out which of the provided answer choices is the correct one. Seems simple, right? Not necessarily.

There are two types of multiple-choice questions. The first is the easy one. It asks a question and provides several answer choices. One of the answer choices is correct and the rest are obviously wrong. Here is an example:

> Who was the fourteenth president of the United States?
> **a.** Walt Disney
> **b.** Tom Cruise
> **c.** Oprah Winfrey
> **d.** Franklin Pierce
> **e.** Homer Simpson

Even if you don't know who was the fourteenth president, you can still answer the question correctly because the wrong answers are obviously wrong. Walt Disney founded the Walt Disney Company, Tom Cruise is an actor, Oprah Winfrey is a talk show host, and Homer Simpson is a cartoon character. Answer choice **c**, Franklin Pierce, is therefore correct.

Unfortunately, the SAT does not include this type of multiple-choice question. Instead, the SAT includes the other type of multiple-choice question. SAT questions include one or more answer choices that *seem correct* but are actually *incorrect*. The test writers include these seemingly correct answer choices to try to trick you into picking the wrong answer.

Let's look at how an SAT writer might write a question about the fourteenth president of the United States:

> Who was the fourteenth president of the United States?
> **a.** George Washington
> **b.** James Buchanan
> **c.** Millard Fillmore
> **d.** Franklin Pierce
> **e.** Abraham Lincoln

This question is much more difficult than the previous question, isn't it? Let's examine what makes it more complicated.

First, all the answer choices are actual presidents. None of the answer choices is obviously wrong. Unless you know exactly which president was the fourteenth, the answer choices don't give you any hints. As a result, you may pick George Washington or Abraham Lincoln because they are two of the best-known presidents. This is exactly what the test writers would want you to do! They included George Washington and Abraham Lincoln because they want you to see a familiar name and assume it's the correct answer.

But what if you know that George Washington was the first president and Abraham Lincoln was the sixteenth president? The question gets even trickier because the other two incorrect answer choices are James Buchanan, the thirteenth president, and Millard Fillmore, the fifteenth president. In other words, unless you happen to know that Franklin Pierce was the fourteenth president, it would be very difficult to figure out he is the correct answer based solely on the answer choices.

In fact, incorrect answer choices are often called **distracters** because they are designed to *distract* you from the correct answer choice.

This is why you should not assume that multiple-choice questions are somehow easier than other types of questions. They can be written to try to trip you up.

But don't worry. There is an important technique that you can use to help make answering multiple-choice questions easier.

▶ Finding Four Incorrect Answer Choices Is the Same as Finding One Correct Answer Choice

Think about it: A multiple-choice question on the SAT has five answer choices. Only one answer choice is correct, which means the other four must be incorrect. You can use this fact to your advantage. Sometimes it's easier to figure out which answer choices are incorrect than to figure out which answer choice is correct.

Here's an exaggerated example:

What is 9,424 × 2,962?

a. 0
b. 10
c. 20
d. 100
e. 27,913,888

Even without doing any calculations, you still know that answer choice **e** is correct because answer choices **a**, **b**, **c**, and **d** are obviously incorrect. Of course, questions on the SAT will not be this easy, but you can still apply this idea to every multiple-choice question on the SAT. Let's see how.

▶ Get Rid of Wrong Answer Choices and Increase Your Luck

Remember that multiple-choice questions on the SAT contain distracters: incorrect answer choices designed to distract you from the correct answer choice. Your job is to get rid of as many of those distracters as you can when answering a question. Even if you can get rid of only one of the five answer choices in a question, you have still increased your chances of answering the question correctly.

Think of it this way: Each SAT question provides five answer choices. If you guess blindly from the five choices, your chances of choosing the correct answer are 1 in 5, or 20%. If you get rid of one answer choice before guessing because you determine that it is incorrect, your chances of choosing the correct answer are 1 in 4, or 25%, because you are choosing from only the four remaining answer choices. If you get rid of two incorrect answer choices before guessing, your chances of choosing the correct answer are 1 in 3, or 33%. Get rid of three incorrect answer choices, and your chances are 1 in 2, or 50%. If you get rid of all four incorrect answer choices, your chances of guessing the correct answer choice are 1 in 1, or 100%! As you can see, each answer choice you eliminate increases your chances of guessing the correct answer.

NUMBER OF DISTRACTERS YOU ELIMINATE	ODDS YOU CAN GUESS THE CORRECT ANSWER
0	1 in 5, or 20%
1	1 in 4, or 25%
2	1 in 3, or 33%
3	1 in 2, or 50%
4	1 in 1, or 100%

Of course, on most SAT questions, you won't be guessing blindly—you'll ideally be able to use your math skills to choose the correct answer—so your chances of picking the correct answer choice are even greater than those listed above after eliminating distracters.

▶ How to Get Rid of Incorrect Answer Choices

Hopefully you are now convinced that getting rid of incorrect answer choices is an important technique to use when answering multiple-choice questions. So how do you do it? Let's look at an example of a question you could see on the SAT.

> The statement below is true.
> All integers in set *A* are odd.
> Which of the following statements must also be true?
> **a.** All even integers are in set *A*.
> **b.** All odd integers are in set *A*.
> **c.** Some integers in set *A* are even.
> **d.** If an integer is even, it is not in set *A*.
> **e.** If an integer is odd, it is not in set *A*.

First, decide what you are looking for: You need to choose which answer choice is true based on the fact that *All integers in set* A *are odd.* This means that the incorrect answer choices are *not* true.

Now follow these steps when answering the question:

1. Evaluate each answer choice one by one following these instructions:
   - If an answer choice is incorrect, cross it out.
   - If you aren't sure if an answer choice is correct or incorrect, leave it alone and go onto the next answer choice.
   - If you find an answer choice that seems correct, circle it and then check the remaining choices to make sure there isn't a better answer.
2. Once you look at all the answer choices, choose the best one from the remaining choices that aren't crossed out.
3. If you can't decide which is the best choice, take your best guess.

Let's try it with the previous question.

Answer choice **a** is *All even integers are in set* A. Let's decide whether this is true. We know that *all integers in set* A *are odd.* This statement means that there are not any even integers in set *A*, so *All even integers are in set* A cannot be true. Cross out answer choice **a**!

Answer choice **b** is *All odd integers are in set* A. Let's decide whether this is true. We know that *all integers in set* A *are odd*, which means that the set could be, for example, {3}, or {1, 3, 5, 7, 9, 11}, or {135, 673, 787}. It describes any set that contains only odd integers, which means that it could also describe a set that contains *all the odd integers.* Therefore, this answer choice may be correct. Let's hold onto it and see how it compares to the other answer choices.

Answer choice **c** is *Some integers in set* A *are even.* We already determined when evaluating answer choice **a** that there are not any even integers in set *A*, so answer choice **c** cannot be true. Cross out answer choice **c**!

Answer choice **d** is *If an integer is even, it is not in set* A. We already determined that there are not any even integers in set *A*, so it seems that *If an integer is even, it is not in set* A is most likely true. This is probably the correct answer. But let's evaluate the last answer choice and then choose the best answer choices from the ones we haven't eliminated.

Answer choice **e** is *If an integer is odd, it is not in set* A. Let's decide whether this is true. We know that *all integers in set* A *are odd*, which means that there is at least one odd integer in set *A* and maybe more. Therefore, answer choice **e** cannot be true. Cross out answer choice **e**!

After evaluating the five answer choices, we are left with answer choices **b** and **d** as the possible correct answer choices. Let's decide which one is better. Answer choice **b** is only possibly true. We know that *all integers in set* A *are odd*, which means that the set contains only odd integers. It could describe a set that contains *all the odd integers*, but it could also describe a set that contains only one odd integer. Answer choice **d**, on the other hand, is always true. If *all integers in set* A *are odd*, then no matter how many integers are in the set, none of them are even. So the statement *If an integer is even, it is not in set* A must be true. It is the better answer choice. Answer choice **d** is correct!

▶ Guessing on Five-Choice Questions: The Long Version

Because five-choice questions provide you with the correct answer as one of their five answer choices, it's possible for you to guess the correct answer even if you don't read the question. You might just get lucky and pick the correct answer.

So should you guess on the SAT if you don't know the answer? Well, it depends. You may have heard that there's a "carelessness penalty" on the SAT. What this means is that careless or random guessing can lower your score. But that doesn't mean you shouldn't guess, because smart guessing can actually work to your advantage and help you earn more points on the exam.

Here's how smart guessing works:

- On the math questions, you get one point for each correct answer. For each question you answer incorrectly, one-fourth of a point is subtracted from your score. If you leave a question blank you are neither rewarded nor penalized.
- On the SAT, all multiple-choice questions have five answer choices. If you guess blindly from among those five choices, you have a one-in-five chance of guessing correctly. That means four times out of five you will probably guess incorrectly. In other words, if there are five questions that you have no clue how to answer, you will probably guess correctly on only one of them and receive one point. You will guess incorrectly on four of them and receive four deductions of one-fourth point each. $1 - \frac{1}{4} - \frac{1}{4} - \frac{1}{4} - \frac{1}{4} = 0$, so if you guess blindly, you will probably neither gain nor lose points in the process.

Why is this important? Well, it means that if you can rule out even one incorrect answer choice on each of the five questions, your odds of guessing correctly improve greatly. So you will receive more points than you will lose by guessing.

In fact, on many SAT questions, it's relatively easy to rule out all but two possible answers. That means you have a 50% chance of being right and receiving one whole point. Of course, you also have a 50% chance of being wrong, but if you choose the wrong answer, you lose only one-fourth point. So for every two questions where you can eliminate all but two answer choices, chances are that you will gain 1 point and lose $\frac{1}{4}$ point, for a gain of $\frac{3}{4}$ points. Therefore, it's to your advantage to guess in these situations!

It's also to your advantage to guess on questions where you can eliminate only one answer choice. If you eliminate one answer choice, you will guess from four choices, so your chances of guessing correctly are 25%. This means that for every four questions where you can eliminate an answer choice, chances are that you will gain 1 point on one of the questions and lose $\frac{1}{4}$ point on the other three questions, for a total gain of $\frac{1}{4}$ point. This may not seem like much, but a $\frac{1}{4}$ point is better than 0 points, which is what you would get if you didn't guess at all.

▶ Guessing on Five-Choice Questions: The Short Version

Okay, who cares about all the reasons you should guess, right? You just want to know when to do it. It's simple:

- If you can eliminate even just one answer choice, you should always guess.
- If you can't eliminate any answer choices, don't guess.

▶ Guessing on Grid-In Questions

The chances of guessing correctly on a grid-in question are so slim that it's usually not worth taking the time to fill in the ovals if you are just guessing blindly. However, you don't lose any points if you answer a grid-in question incorrectly, so if you have some kind of attempt at an answer, fill it in!

To summarize:

- If you've figured out a solution to the problem—even if you think it might be incorrect—fill in the answer.
- If you don't have a clue about how to answer the question, don't bother guessing.

▶ Other Important Strategies

Read the Questions Carefully and Know What the Question Is Asking You to Do

Many students read questions too quickly and don't understand what exactly they should answer before examining the answer choices. Questions are often written to trick students into choosing an incorrect answer choice based on misunderstanding the question. So always read questions carefully. When you finish reading the question, make a note of what you should look for in the answer choices. For example, it might be, "I need to determine the *y*-intercept of the line when its slope is 4" or "I need to determine the area of the unshaded region in the figure."

If You Are Stuck on a Question after 30 Seconds, Move On to the Next Question

You have 25 minutes to answer questions in each of two math sections and 20 minutes to answer questions in the third math section. In all, you must answer 65 questions in 70 minutes. That means you have about a minute per question. On many questions, less than a minute is all you will need. On others, you'll wish you had much longer than a minute. But don't worry! The SAT is designed to be too complex to finish. Therefore, do not waste time on a difficult question until you have completed the questions you know how to solve. If you can't figure out how to solve a question in 30 seconds or so and you are just staring at the page, move on to the next question. However, if you feel you are making good progress on a question, finish answering it, even if it takes you a minute or a little more.

Start with Question 1, Not Question 25

The SAT math questions can be rated from 1–5 in level of difficulty, with 1 being the easiest and 5 being the most difficult. The following is an example of how questions of varying difficulty are typically distributed in one section of a typical SAT. (**Note:** The distribution of questions on your test will vary. This is only an example.)

1. 1	**8.** 2	**15.** 3	**22.** 3
2. 1	**9.** 3	**16.** 5	**23.** 5
3. 1	**10.** 2	**17.** 4	**24.** 5
4. 1	**11.** 3	**18.** 4	**25.** 5
5. 2	**12.** 3	**19.** 4	
6. 2	**13.** 3	**20.** 4	
7. 1	**14.** 3	**21.** 4	

From this list, you can see how important it is to complete the first fifteen questions of one section before you get bogged down in the more difficult questions that follow. Because all the questions are worth the same amount, you should be sure to get the easiest questions correct. So make sure that you answer the first 15 questions well! These are typically the questions that are easiest to answer correctly. Then, after you are satisfied with the first fifteen questions, answer the rest. If you can't figure out how to solve a question after 30 seconds, move onto the next one. Spend the most time on questions that you think you can solve, not the questions that you are confused about.

Pace Yourself

We just told you that you have about a minute to answer each question. But this doesn't mean you should rush! There's a big difference between rushing and pacing yourself so you don't waste time.

Many students rush when they take the SAT. They worry they won't have time to answer all the questions. But here's some important advice: It is better to answer most questions correctly and leave some blank at the end than to answer every question but make a lot of careless mistakes.

As we said, on average you have a little over a minute to answer each math question on the SAT. Some questions will require less time than that. Others will require more. A minute may not seem like a long time to answer a question, but it usually is. As an experiment, find a clock and watch the second hand move as you sit silently for one minute. You'll see that a minute lasts longer than you think.

So how do you make sure you keep on a good pace? The best strategy is to work on one question at a time. Don't worry about any future questions or any previous questions you had trouble with. Focus all your attention on the present question. Start with Question 1. If you determine an answer in less than a minute, mark it and move to Question 2. If you can't decide on an answer in less than a minute, take your best guess from the answer choices you haven't eliminated, circle the question, and move on. If you have time at the end of the section, you can look at the question again. But in the meantime, forget about it. Concentrate on Question 2.

Follow this strategy throughout each section:

1. Focus.
2. Mark an answer.
3. Circle the question if you want to go back to it later.
4. Then, move on to the next question.

Hopefully you will be able to answer the first several easier questions in much less than a minute. This will give you extra time to spend on the more difficult questions at the end of the section. But remember: Easier questions are worth the same as the more difficult questions. It's better to get all the easier questions right and all the more difficult questions wrong than to get a lot of the easier questions wrong because you were too worried about the more difficult questions.

Don't Be Afraid to Write in Your Test Booklet

The test scorers will not evaluate your test booklet, so feel free to write in it in any way that will help you during the exam. For example, mark each question that you don't answer so that you can go back to it later. Then, if you have extra time at the end of the section, you can easily find the questions that need extra attention. It is also helpful to cross out the answer choices that you have eliminated as you answer each question.

On Some Questions, It May Be Best to Substitute in an Answer Choice

Sometimes it is quicker to pick an answer choice and check to see if it works as a solution then to try to find the solution and then choose an answer choice.

Example

The average of 8, 12, 7, and a is 10. What is the value of a?

a. 10
b. 13
c. 19
d. 20
e. 27

One way to solve this question is with algebra. Because the average of four numbers is determined by the sum of the four numbers divided by 4, you could write the following equation and solve for a:

$$\frac{8 + 12 + 7 + a}{4} = 10$$

$$\frac{8 + 12 + 7 + a}{4} \times 4 = 10 \times 4$$

$$8 + 12 + 7 + a = 40$$

$$27 + a = 40$$

$$27 + a - 27 = 40 - 27$$

$$a = 13$$

However, you can also solve this problem without algebra. You can write the expression $\frac{8 + 12 + 7 + a}{4}$ and just substitute each answer choice for a until you find one that makes the expression equal to 10.

Tip: When you substitute an answer choice, always start with answer choice **c**. Answer choices are ordered from least to greatest, so answer choice **c** will be the middle number. Then you can adjust the outcome to the problem as needed by choosing answer choice **b** or **d** next, depending on whether you need a larger or smaller answer.

Let's see how it works:

Answer choice **c**: $\frac{8 + 12 + 7 + 19}{4} = \frac{45}{4}$, which is greater than 10. Therefore, we need a small answer choice. Try choice **b** next:
Answer choice **b**: $\frac{8 + 12 + 7 + 13}{4} = \frac{40}{4} = 10$
There! You found the answer. The variable a must be 13. Therefore answer choice **b** is correct.

Of course, solving this problem with algebra is fine, too. But you may find that substitution is quicker and/or easier. So if a question asks you to solve for a variable, consider using substitution.

Convert All Units of Measurement to the Same Units Used in the Answer Choices before Solving the Problem

If a question involves units of measurement, be sure to convert all units in the question to the units used in the answer choices before you solve the problem. If you wait to convert units later, you may forget to do it and will choose an incorrect answer. If you make the conversions at the start of the problem, you won't have to worry about them later. You can then focus on finding an answer instead of worrying about what units the answer should be in. For example, if the answer choices of a word problem are in *feet* but the problem includes measurements in *inches,* convert all measurements to feet before making any calculations.

Draw Pictures When Solving Word Problems if Needed

Pictures are usually helpful when a word problem doesn't have one, especially when the problem is dealing with geometry. Also, many students are better at solving problems when they see a visual representation. But don't waste time making any drawings too elaborate. A simple drawing, labeled correctly, is usually all you need.

Avoid Lengthy Calculations

It is seldom, if ever, necessary to spend a great deal of time doing calculations. The SAT is a test of mathematical concepts, not calculations. If you find yourself doing a very complex, lengthy calculation—stop! Either you are not solving the problem correctly or you are missing an easier method.

Don't Overuse Your Calculator

Because not every student will have a calculator, the SAT does not include questions that require you to use one. As a result, calculations are generally not complex. So don't make your solutions too complicated simply because you have a calculator handy. Use your calculator sparingly. It will not help you much on the SAT.

Fill in Answer Ovals Carefully and Completely

The Math sections of the SAT are scored by computer. All the computer cares about is whether the correct answer oval is filled in. So fill in your answer ovals neatly! Make sure each oval is filled in completely and that there are no stray marks on the answer sheet. You don't want to lose any points because the computer can't understand which oval you filled in.

Mark Your Answer Sheet Carefully

This may seem obvious, but you must be careful that you fill in the correct answer oval on the answer sheet for each question. Answer sheets can be confusing—so many lines of ovals. So always double-check that you are filling in the correct oval under the correct question number. If you know the correct answer to question 12 but you fill it in under question 11 on the answer sheet, it will be marked as incorrect!

If You Have Time, Double-Check Your Answers

If you finish a section early, use the extra time to double-check your answers. It is common to make careless errors on timed tests, so even if you think you answered every question correctly, it won't hurt to check your answers again. You should also check your answer sheet and make sure that you have filled in your answers clearly and that you haven't filled in more than one oval for any question.

▶ . . . And Don't Forget to Practice!

The strategies in this chapter will definitely help you on the five-choice questions, but simply reading the strategies is not enough. For maximum benefit, you must practice, practice, and practice. So apply these strategies to all the practice questions in this book. The more comfortable you become in answering SAT questions using these strategies, the better you will perform on the test!

▶ Before the Test: Your Final Preparation

Your routine in the last week before the test should vary from your study routine of the preceding weeks.

The Final Week

Saturday morning, one week before you take the SAT, take a final practice test. Then use your next few days to wrap up any loose ends. This week is also the time to read back over your notes on test-taking tips and techniques.

However, it's a good idea to actually cut back on your study schedule in the final week. The natural tendency is to cram before a big test. Maybe this strategy has worked for you with other exams, but it's not a good idea with the SAT. Also, cramming tends to raise your anxiety level, and your brain doesn't do its best work when you're anxious. Anxiety is your enemy when it comes to test taking. It's also your enemy when it comes to restful sleep, and it's extremely important that you be well rested and relaxed on test day.

During the last week before the exam, make sure you know where you're taking the test. If it's an unfamiliar location, drive there so you will know how long it takes to get there, how long it takes to park, and how long to walk from the car to the building where you will take the SAT. This way you can avoid a last minute rush to the test.

Be sure you get adequate exercise during this last week. Exercise will help you sleep soundly and will help rid your body and mind of the effects of anxiety. Don't tackle any new physical skills, though, or overdo any old ones. You don't want to be sore and uncomfortable on test day!

Check to see that your test admission ticket and your personal identification are in order and easily located. Sharpen your pencils. Buy new batteries for your calculator and put them in.

The Day Before

It's the day before the SAT. Here are some dos and don'ts:

DOs

- Relax!
- Find something fun to do the night before—watch a good movie, have dinner with a friend, read a good book.
- Get some light exercise. Walk, dance, swim.
- Get together everything you need for the test: admission ticket, ID, #2 pencils, calculator, watch, bottle of water, and snacks.
- Go to bed early. Get a good night's sleep.

DON'Ts

- Do not study. You've prepared. Now relax.
- Don't party. Keep it low key.
- Don't eat anything unusual or adventurous—save it!
- Don't try any unusual or adventurous activity—save it!
- Don't allow yourself to get into an emotional exchange with anyone—a parent, a sibling, a friend, or a significant other. If someone starts an argument, remind him or her you have an SAT to take and need to postpone the discussion so you can focus on the exam.

Test Day

On the day of the test, get up early enough to allow yourself extra time to get ready. Set your alarm and ask a family member or friend to make sure you are up.

Eat a light, healthy breakfast, even if you usually don't eat in the morning. If you don't normally drink coffee, don't do it today. If you do normally have coffee, have only one cup. More than one cup may make you jittery. If you plan to take snacks for the break, take something healthy and easy to manage. Nuts and raisins are a great source of long-lasting energy. Stay away from cookies and candy during the exam. Remember to take water.

Give yourself plenty of time to get to the test site and avoid a last-minute rush. Plan to get to the test room ten to fifteen minutes early.

During the exam, check periodically (every five to ten questions) to make sure you are transposing your answers to the answer sheet correctly. Look at the question number, then check your answer sheet to see that you are marking the oval by that question number.

If you find yourself getting anxious during the test, remember to breathe. Remember that you have worked hard to prepare for this day. You are ready.

CHAPTER

5

Numbers and Operations Review

This chapter reviews key concepts of numbers and operations that you need to know for the SAT. Throughout the chapter are sample questions in the style of SAT questions. Each sample SAT question is followed by an explanation of the correct answer.

▶ Real Numbers

All numbers on the SAT are real numbers. Real numbers include the following sets:

- **Whole numbers** are also known as counting numbers.
  0, 1, 2, 3, 4, 5, 6, . . .
- **Integers** are positive and negative whole numbers and the number zero.
  . . . –3, –2, –1, 0, 1, 2, 3 . . .
- **Rational numbers** are all numbers that can be written as fractions, terminating decimals, and repeating decimals. Rational numbers include integers.
  $\frac{3}{4}$ $\frac{2}{1}$ 0.25 0.38658 $0.\overline{666}$
- **Irrational numbers** are numbers that cannot be expressed as terminating or repeating decimals.
  π $\sqrt{2}$ 1.6066951524 . . .

Practice Question

The number –16 belongs in which of the following sets of numbers?

a. rational numbers only
b. whole numbers and integers
c. whole numbers, integers, and rational numbers
d. integers and rational numbers
e. integers only

Answer

d. –16 is an integer because it is a negative whole number. It is also a rational number because it can be written as a fraction. All integers are also rational numbers. It is not a whole number because negative numbers are not whole numbers.

▶ Comparison Symbols

The following table shows the various comparison symbols used on the SAT.

SYMBOL	MEANING	EXAMPLE
$=$	is equal to	$3 = 3$
$\neq$	is not equal to	$7 \neq 6$
$>$	is greater than	$5 > 4$
$\geq$	is greater than or equal to	$x \geq 2$ (x can be 2 or any number greater than 2)
$<$	is less than	$1 < 2$
$\leq$	is less than or equal to	$x \leq 8$ (x can be 8 or any number less than 8)

Practice Question

If $a > 37$, which of the following is a possible value of a?

a. –43
b. –37
c. 35
d. 37
e. 41

Answer

e. $a > 37$ means that a is greater than 37. Only 41 is greater than 37.

Symbols of Multiplication

A **factor** is a number that is multiplied. A **product** is the result of multiplication.

$7 \times 8 = 56$. 7 and 8 are *factors.* 56 is the *product.*

You can represent multiplication in the following ways:

- A multiplication sign or a dot between factors indicates multiplication:
  $7 \times 8 = 56$ $\quad$ $7 \cdot 8 = 56$
- Parentheses around a factor indicate multiplication:
  $(7)8 = 56$ $\quad$ $7(8) = 56$ $\quad$ $(7)(8) = 56$
- Multiplication is also indicated when a number is placed next to a variable:
  $7a = 7 \times a$

Practice Question

If $n = (8 - 5)$, what is the value of $6n$?

a. 2
b. 3
c. 6
d. 9
e. 18

Answer

e. $6n$ means $6 \times n$, so $6n = 6 \times (8 - 5) = 6 \times 3 = 18$.

Like Terms

A **variable** is a letter that represents an unknown number. Variables are used in equations, formulas, and mathematical rules.

A number placed next to a variable is the **coefficient** of the variable:

$9d$	9 is the coefficient to the variable d.
$12xy$	12 is the coefficient to both variables, x and y.

If two or more terms contain exactly the same variables, they are considered **like terms:**

$-4x$, $7x$, $24x$, and $156x$ are all like terms.
$-8ab$, $10ab$, $45ab$, and $217ab$ are all like terms.

Variables with different exponents are **not** like terms. For example, $5x^3y$ and $2xy^3$ are not like terms. In the first term, the x is cubed, and in the second term, it is the y that is cubed.

You can **combine like terms** by grouping like terms together using mathematical operations:

$3x + 9x = 12x$ $\quad$ $17a - 6a = 11a$

Practice Question

$4x^2y + 5y + 7xy + 8x + 9xy + 6y + 3xy^2$

Which of the following is equal to the expression above?

a. $4x^2y + 3xy^2 + 16xy + 8x + 11y$
b. $7x^2y + 16xy + 8x + 11y$
c. $7x^2y^2 + 16xy + 8x + 11y$
d. $4x^2y + 3xy^2 + 35xy$
e. $23x^4y^4 + 8x + 11y$

Answer

a. Only like terms can be combined in an expression. $7xy$ and $9xy$ are like terms because they share the same variables. They combine to $16xy$. $5y$ and $6y$ are also like terms. They combine to $11y$. $4x^2y$ and $3xy^2$ are not like terms because their variables have different exponents. In one term, the x is squared, and in the other, it's not. Also, in one term, the y is squared and in the other it's not. Variables must have the exact same exponents to be considered like terms.

▶ Properties of Addition and Multiplication

- **Commutative Property of Addition**. When using addition, the order of the addends does not affect the sum:
  $a + b = b + a$ $\quad$ $7 + 3 = 3 + 7$
- **Commutative Property of Multiplication**. When using multiplication, the order of the factors does not affect the product:
  $a \times b = b \times a$ $\quad$ $6 \times 4 = 4 \times 6$
- **Associative Property of Addition.** When adding three or more addends, the grouping of the addends does not affect the sum.
  $a + (b + c) = (a + b) + c$ $\quad$ $4 + (5 + 6) = (4 + 5) + 6$
- **Associative Property of Multiplication.** When multiplying three or more factors, the grouping of the factors does not affect the product.
  $5(ab) = (5a)b$ $\quad$ $(7 \times 8) \times 9 = 7 \times (8 \times 9)$
- **Distributive Property**. When multiplying a sum (or a difference) by a third number, you can multiply each of the first two numbers by the third number and then add (or subtract) the products.
  $7(a + b) = 7a + 7b$ $\quad$ $9(a - b) = 9a - 9b$
  $3(4 + 5) = 12 + 15$ $\quad$ $2(3 - 4) = 6 - 8$

Practice Question

Which equation illustrates the commutative property of multiplication?

a. $7(\frac{8}{9} + \frac{3}{10}) = (7 \times \frac{8}{9}) + (7 \times \frac{3}{10})$
b. $(4.5 \times 0.32) \times 9 = 9 \times (4.5 \times 0.32)$
c. $12(0.65 \times 9.3) = (12 \times 0.65) \times (12 \times 9.3)$
d. $(9.04 \times 1.7) \times 2.2 = 9.04 \times (1.7 \times 2.2)$
e. $5 \times (\frac{3}{7} \times \frac{4}{9}) = (5 \times \frac{3}{7}) \times \frac{4}{9}$

Answer

b. Answer choices **a** and **c** show the distributive property. Answer choices **d** and **e** show the associative property. Answer choice **b** is correct because it represents that you can change the order of the terms you are multiplying without affecting the product.

Order of Operations

You must follow a specific order when calculating multiple operations:

Parentheses: First, perform all operations within parentheses.
Exponents: Next evaluate exponents.
Multiply/**D**ivide: Then work from left to right in your multiplication and division.
Add/**S**ubtract: Last, work from left to right in your addition and subtraction.

You can remember the correct order using the acronym **PEMDAS** or the mnemonic *Please Excuse My Dear Aunt Sally*.

Example

$8 + 4 \times (3 + 1)^2$

$= 8 + 4 \times (4)^2$	**P**arentheses
$= 8 + 4 \times 16$	**E**xponents
$= 8 + 64$	**M**ultiplication (and **D**ivision)
$= 72$	**A**ddition (and **S**ubtraction)

Practice Question

$3 \times (49 - 16) + 5 \times (2 + 3^2) - (6 - 4)^2$

What is the value of the expression above?

a. 146
b. 150
c. 164
d. 220
e. 259

Answer

b. Following the order of operations, the expression should be simplified as follows:

$3 \times (49 - 16) + 5 \times 3\,(2 + 3^2) - (6 - 4)^2$

$3 \times (33) + 5 \times (2 + 9) - (2)^2$

$3 \times (33) + 5 \times (11) - 4$

$[3 \times (33)] + [5 \times (11)] - 4$

$99 + 55 - 4$

$= 150$

▶ Powers and Roots

Exponents

An **exponent** tells you how many times a number, the **base,** is a factor in the product.

$3^5 = 3 \times 3 \times 3 \times 3 \times 3 = 243$ 3 is the *base.* 5 is the *exponent.*

Exponents can also be used with variables. You can substitute for the variables when values are provided.

b^n The "b" represents a number that will be a factor to itself "n" times.
If $b = 4$ and $n = 3$, then $b^n = 4^3 = 4 \times 4 \times 4 = 64$.

Practice Question

Which of the following is equivalent to 7^8?

a. $7 \times 7 \times 7 \times 7 \times 7 \times 7$
b. $7 \times 7 \times 7 \times 7 \times 7 \times 7 \times 7$
c. $8 \times 8 \times 8 \times 8 \times 8 \times 8 \times 8$
d. $7 \times 7 \times 7 \times 7 \times 7 \times 7 \times 7 \times 7$
e. $7 \times 8 \times 7 \times 8$

Answer

d. 7 is the base. 8 is the exponent. Therefore, 7 is multiplied 8 times.

Laws of Exponents

- Any base to the zero power equals 1.
  $(12xy)^0 = 1$ $80^0 = 1$ $8{,}345{,}832^0 = 1$
- When multiplying identical bases, keep the same base and add the exponents:
  $b^m \times b^n = b^{m+n}$

Examples

$9^5 \times 9^6 = 9^{5+6} = 9^{11}$ $\quad$ $a^2 \times a^3 \times a^5 = a^{2+3+5} = a^{10}$

- When dividing identical bases, keep the same base and subtract the exponents:

$b^m \div b^n = b^{m-n}$ $\quad$ $\frac{b^m}{b^n} = b^{m-n}$

Examples

$6^5 \div 6^3 = 6^{5-3} = 6^2$ $\quad$ $\frac{a^9}{a^4} = a^{9-4} = a^5$

- If an exponent appears outside of parentheses, multiply any exponents inside the parentheses by the exponent outside the parentheses.

$(b^m)^n = b^{m \times n}$

Examples

$(4^3)^8 = 4^{3 \times 8} = 4^{24}$ $\quad$ $(j^4 \times k^2)^3 = j^{4 \times 3} \times k^{2 \times 3} = j^{12} \times k^6$

Practice Question

Which of the following is equivalent to 6^{12}?

a. $(6^6)^6$

b. $6^2 + 6^5 + 6^5$

c. $6^3 \times 6^2 \times 6^7$

d. $\frac{18^{15}}{3^3}$

e. $\frac{6^4}{6^3}$

Answer

c. Answer choice **a** is incorrect because $(6^6)^6 = 6^{36}$. Answer choice **b** is incorrect because exponents don't combine in addition problems. Answer choice **d** is incorrect because $\frac{bm}{bn} = b^{m-n}$ applies only when the base in the numerator and denominator are the same. Answer choice **e** is incorrect because you must subtract the exponents in a division problem, not multiply them. Answer choice **c** is correct: $6^3 \times 6^2 \times 6^7 = 6^{3+2+7} = 6^{12}$.

▶ Squares and Square Roots

The **square** of a number is the product of a number and itself. For example, the number 25 is the **square** of the number 5 because $5 \times 5 = 25$. The square of a number is represented by the number raised to a power of 2:

$a^2 = a \times a$ $\quad$ $5^2 = 5 \times 5 = 25$

The **square root** of a number is one of the equal factors whose product is the square. For example, 5 is the square root of the number 25 because $5 \times 5 = 25$. The symbol for square root is $\sqrt{\ }$. This symbol is called the **radical.** The number inside of the radical is called the **radicand.**

$\sqrt{36} = 6$ because $6^2 = 36$ 36 is the square of 6, so 6 is the square root of 36.

Practice Question

Which of the following is equivalent to $\sqrt{196}$?

a. 13
b. 14
c. 15
d. 16
e. 17

Answer

b. $\sqrt{196} = 14$ because $14 \times 14 = 196$.

Perfect Squares

The square root of a number might not be a whole number. For example, there is not a whole number that can be multiplied by itself to equal 8. $\sqrt{8} = 2.8284271\ldots.$

A whole number is a **perfect square** if its square root is also a whole number:

1 is a perfect square because $\sqrt{1} = 1$
4 is a perfect square because $\sqrt{4} = 2$
9 is a perfect square because $\sqrt{9} = 3$
16 is a perfect square because $\sqrt{16} = 4$
25 is a perfect square because $\sqrt{25} = 5$
36 is a perfect square because $\sqrt{36} = 6$
49 is a perfect square because $\sqrt{49} = 7$

Practice Question

Which of the following is a perfect square?

a. 72
b. 78
c. 80
d. 81
e. 88

Answer

d. Answer choices **a**, **b**, **c**, and **e** are incorrect because they are not perfect squares. The square root of a perfect square is a whole number; $\sqrt{72} \approx 8.485$; $\sqrt{78} \approx 8.832$; $\sqrt{80} \approx 8.944$; $\sqrt{88} \approx 9.381$; 81 is a perfect square because $\sqrt{81} = 9$.

Properties of Square Root Radicals

The product of the square roots of two numbers is the same as the square root of their product.

$\sqrt{a} \times \sqrt{b} = \sqrt{a \times b}$ $\quad$ $\sqrt{7} \times \sqrt{3} = \sqrt{7 \times 3} = \sqrt{21}$

- The quotient of the square roots of two numbers is the square root of the quotient of the two numbers.

$\frac{\sqrt{a}}{\sqrt{b}} = \sqrt{\frac{a}{b}}$, where $b \neq 0$ $\quad$ $\frac{\sqrt{24}}{\sqrt{8}} = \sqrt{\frac{24}{8}} = \sqrt{3}$

- The square of a square root radical is the radicand.

$(\sqrt{N})^2 = N$ $\quad$ $(\sqrt{4})^2 = \sqrt{4} \times \sqrt{4} = \sqrt{16} = 4$

- When adding or subtracting radicals with the same radicand, add or subtract only the coefficients. Keep the radicand the same.

$a\sqrt{b} + c\sqrt{b} = (a + c)\sqrt{b}$ $\quad$ $4\sqrt{7} + 6\sqrt{7} = (4 + 6)\sqrt{7} = 10\sqrt{7}$

- You cannot combine radicals with different radicands using addition or subtraction.

$\sqrt{a} + \sqrt{b} \neq \sqrt{a + b}$ $\quad$ $\sqrt{2} + \sqrt{3} \neq \sqrt{5}$

- To simplify a square root radical, write the radicand as the product of two factors, with one number being the largest perfect square factor. Then write the radical over each factor and simplify.

$\sqrt{8} = \sqrt{4} \times \sqrt{2} = 2 \times \sqrt{2} = 2\sqrt{2}$ $\quad$ $\sqrt{27} = \sqrt{9} \times \sqrt{3} = 3 \times \sqrt{3} = 3\sqrt{3}$

Practice Question

Which of the following is equivalent to $2\sqrt{6}$?

a. $2\sqrt{3} \times \sqrt{3}$

b. $\sqrt{24}$

c. $\frac{2\sqrt{9}}{\sqrt{3}}$

d. $2\sqrt{4} + 2\sqrt{2}$

e. $\sqrt{72}$

Answer

b. Answer choice **a** is incorrect because $2\sqrt{3} \times \sqrt{3} = 2\sqrt{9}$. Answer choice **c** is incorrect because $\frac{2\sqrt{9}}{\sqrt{3}} = 2\sqrt{3}$. Answer choice **d** is incorrect because you cannot combine radicals with different radicands using addition or subtraction. Answer choice **e** is incorrect because $\sqrt{72} = \sqrt{2 \times 36} = 6\sqrt{2}$. Answer choice **b** is correct because $\sqrt{24} = \sqrt{6 \times 4} = 2\sqrt{6}$.

▶ Negative Exponents

Negative exponents are the opposite of positive exponents. Therefore, because positive exponents tell you how many of the base to *multiply* together, negative exponents tell you how many of the base to *divide*.

$a^{-n} = \frac{1}{a^n}$ $3^{-2} = \frac{1}{3^2} = \frac{1}{3 \times 3} = \frac{1}{9}$ $-5^{-3} = -\frac{1}{5^3} = -\frac{1}{5 \times 5 \times 5} = -\frac{1}{125}$

Practice Question

Which of the following is equivalent to -6^{-4}?

a. $-1{,}296$

b. $-\frac{6}{1{,}296}$

c. $-\frac{1}{1{,}296}$

d. $\frac{1}{1{,}296}$

e. 1,296

Answer

c. $-6^{-4} = -\frac{1}{6^4} = -\frac{1}{6 \times 6 \times 6 \times 6} = -\frac{1}{1{,}296}$

▶ Rational Exponents

Rational numbers are numbers that can be written as fractions (and decimals and repeating decimals). Similarly, numbers raised to rational exponents are numbers raised to fractional powers:

$4^{\frac{1}{2}}$ $25^{\frac{1}{2}}$ $8^{\frac{1}{3}}$ $3^{\frac{2}{3}}$

For a number with a fractional exponent, the numerator of the exponent tells you the power to raise the number to, and the denominator of the exponent tells you the root you take.

$4^{\frac{1}{2}} = \sqrt{4^1} = \sqrt{4} = 2$

The numerator is 1, so raise 4 to a power of 1. The denominator is 2, so take the square root.

$25^{\frac{1}{2}} = \sqrt{25^1} = \sqrt{25} = 5$

The numerator is 1, so raise 25 to a power of 1. The denominator is 2, so take the square root.

$8^{\frac{1}{3}} = \sqrt[3]{8^1} = \sqrt[3]{8} = 2$

The numerator is 1, so raise 8 to a power of 1. The denominator is 3, so take the cube root.

$$3^{\frac{2}{3}} = \sqrt[3]{3^2} = \sqrt[3]{9}$$

The numerator is 2, so raise 3 to a power of 2. The denominator is 3, so take the cube root.

Practice Question

Which of the following is equivalent to $8^{\frac{2}{3}}$?

a. $\sqrt[3]{4}$
b. $\sqrt[3]{8}$
c. $\sqrt[3]{16}$
d. $\sqrt[3]{64}$
e. $\sqrt{512}$

Answer

d. In the exponent of $8^{\frac{2}{3}}$, the numerator is 2, so raise 8 to a power of 2. The denominator is 3, so take the cube root; $\sqrt[3]{8^2} = \sqrt[3]{64}$.

▶ Divisibility and Factors

Like multiplication, division can be represented in different ways. In the following examples, 3 is the **divisor** and 12 is the **dividend.** The result, 4, is the **quotient**.

$12 \div 3 = 4$ $\qquad 3\overline{)12} = 4$ $\qquad \frac{12}{3} = 4$

Practice Question

In which of the following equations is the divisor 15?

a. $\frac{15}{5} = 3$
b. $\frac{60}{15} = 4$
c. $15 \div 3 = 5$
d. $45 \div 3 = 15$
e. $10\overline{)150} = 15$

Answer

b. The divisor is the number that divides *into* the dividend to find the quotient. In answer choices **a** and **c**, 15 is the dividend. In answer choices **d** and **e**, 15 is the quotient. Only in answer choice **b** is 15 the divisor.

Odd and Even Numbers

An **even** number is a number that can be divided by the number 2 to result in a whole number. Even numbers have a 2, 4, 6, 8, or 0 in the ones place.

2 34 86 1,018 6,987,120

Consecutive even numbers differ by two:

2, 4, 6, 8, 10, 12, 14 . . .

An **odd** number cannot be divided evenly by the number 2 to result in a whole number. Odd numbers have a 1, 3, 5, 7, or 9 in the ones place.

1 13 95 2,827 7,820,289

Consecutive odd numbers differ by two:

1, 3, 5, 7, 9, 11, 13 . . .

Even and odd numbers behave consistently when added or multiplied:

$even + even = even$	and	$even \times even = even$
$odd + odd = even$	and	$odd \times odd = odd$
$odd + even = odd$	and	$even \times odd = even$

Practice Question

Which of the following situations must result in an odd number?

a. even number + even number
b. odd number × odd number
c. odd number + 1
d. odd number + odd number
e. $\frac{\text{even number}}{2}$

Answer

b. **a**, **c**, and **d** definitely yield even numbers; **e** could yield either an even or an odd number. The product of two odd numbers (**b**) is an odd number.

Dividing by Zero

Dividing by zero is impossible. Therefore, the denominator of a fraction can never be zero. Remember this fact when working with fractions.

Example

$\frac{5}{n-4}$ We know that $n \neq 4$ because the denominator cannot be 0.

Factors

Factors of a number are whole numbers that, when divided into the original number, result in a quotient that is a whole number.

Example

The factors of 18 are 1, 2, 3, 6, 9, and 18 because these are the only whole numbers that divide evenly into 18.

The **common factors** of two or more numbers are the factors that the numbers have in common. The **greatest common factor** of two or more numbers is the largest of all the common factors. Determining the greatest common factor is useful for reducing fractions.

Examples

The *factors* of 28 are 1, 2, 4, 7, 14, and 28.
The *factors* of 21 are 1, 3, 7, and 21.
The *common factors* of 28 and 21 are therefore 1 and 7 because they are factors of both 28 and 21.
The *greatest common factor* of 28 and 21 is therefore 7. It is the largest factor shared by 28 and 21.

Practice Question

What are the common factors of 48 and 36?

a. 1, 2, and 3
b. 1, 2, 3, and 6
c. 1, 2, 3, 6, and 12
d. 1, 2, 3, 6, 8, and 12
e. 1, 2, 3, 4, 6, 8, and 12

Answer

c. The factors of 48 are 1, 2, 3, 6, 8, 12, 24, and 48. The factors of 36 are 1, 2, 3, 6, 12, 18, and 36. Therefore, their common factors—the factors they share—are 1, 2, 3, 6, and 12.

▶ Multiples

Any number that can be obtained by multiplying a number x by a whole number is called a **multiple** of x.

Examples

Multiples of x include $1x$, $2x$, $3x$, $4x$, $5x$, $6x$, $7x$, $8x$ . . .
Multiples of 5 include 5, 10, 15, 20, 25, 30, 35, 40 . . .
Multiples of 8 include 8, 16, 24, 32, 40, 48, 56, 64 . . .

The **common multiples** of two or more numbers are the multiples that the numbers have in common. The **least common multiple** of two or more numbers is the smallest of all the common multiples. The least common multiple, or LCM, is used when performing various operations with fractions.

Examples

Multiples of 10 include 10, 20, 30, 40, 50, 60, 70, 80, 90 . . .

Multiples of 15 include 15, 30, 45, 60, 75, 90, 105 . . .

Some *common multiples* of 10 and 15 are therefore 30, 60, and 90 because they are multiples of both 10 and 15.

The *least common multiple* of 10 and 15 is therefore 30. It is the smallest of the multiples shared by 10 and 15.

▶ Prime and Composite Numbers

A positive integer that is greater than the number 1 is either prime or composite, but not both.

- A **prime number** has only itself and the number 1 as factors:
  2, 3, 5, 7, 11, 13, 17, 19, 23 . . .
- A **composite** number is a number that has more than two factors:
  4, 6, 8, 9, 10, 12, 14, 15, 16 . . .
- The number 1 is neither prime nor composite.

Practice Question

n is a prime number and

$n > 2$

What must be true about n?

a. $n = 3$

b. $n = 4$

c. n is a negative number

d. n is an even number

e. n is an odd number

Answer

e. All prime numbers greater than 2 are odd. They cannot be even because all even numbers are divisible by at least themselves *and* the number 2, which means they have at least two factors and are therefore composite, not prime. Thus, answer choices **b** and **d** are incorrect. Answer choice **a** is incorrect because, although n could equal 3, it does not necessarily equal 3. Answer choice **c** is incorrect because $n > 2$.

▶ Prime Factorization

Prime factorization is a process of breaking down factors into prime numbers.

Example

Let's determine the prime factorization of 18.

Begin by writing 18 as the product of two factors:

$18 = 9 \times 2$

Next break down those factors into smaller factors:

9 can be written as 3×3, so $18 = 9 \times 2 = 3 \times 3 \times 2$.

The numbers 3, 3, and 2 are all prime, so we have determined that the prime factorization of 18 is $3 \times 3 \times 2$.

We could have also found the prime factorization of 18 by writing the product of 18 as 3×6:

6 can be written as 3×2, so $18 = 6 \times 3 = 3 \times 3 \times 2$.

Thus, the prime factorization of 18 is $3 \times 3 \times 2$.

Note: Whatever the road one takes to the factorization of a number, the answer is always the same.

Practice Question

$2 \times 2 \times 2 \times 5$ is the prime factorization of which number?

a. 10
b. 11
c. 20
d. 40
e. 80

Answer

d. There are two ways to answer this question. You could find the prime factorization of each answer choice, or you could simply multiply the prime factors together. The second method is faster: $2 \times 2 \times 2 \times 5 = 4 \times 2 \times 5 = 8 \times 5 = 40$.

▶ Number Lines and Signed Numbers

On a number line, *less than 0* is to the *left* of 0 and *greater than 0* is to the *right* of 0.

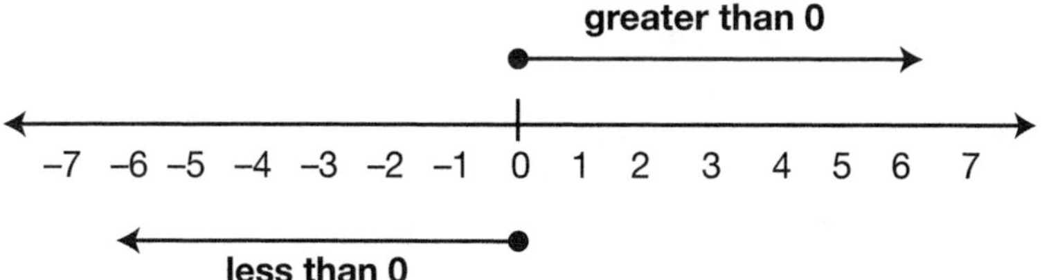

Negative numbers are the opposites of positive numbers.

Examples

5 is five to the *right* of zero.

-5 is five to the *left* of zero.

If a number is *less than* another number, it is farther to the left on the number line.

Example

-4 is to the left of -2, so $-4 < -2$.

If a number is *greater than* another number, it is farther to the right on the number line.

Example

3 is to the right of -1, so $3 > -1$.

A positive number is always greater than a negative number. A negative number is always less than a positive number.

Examples

2 is greater than $-3,675$.

$-25,812$ is less than 3.

As a shortcut to avoiding confusion when comparing two negative numbers, remember the following rules:

When a and b are positive, if $a > b$, then $-a < -b$.

When a and b are positive, if $a < b$, then $-a > -b$.

Examples

If $8 > 6$, then $-6 > -8$. (8 is to the right of 6 on the number line. Therefore, -8 is to the left of -6 on the number line.)

If $132 < 267$, then $-132 > -267$. (132 is to the left of 267 on the number line. Therefore, -132 is to the right of -267 on the number line.)

Practice Question

Which of the following statements is true?

a. $-25 > -24$
b. $-48 > 16$
c. $14 > 17$
d. $-22 > 19$
e. $-37 > -62$

Answer

e. $-37 > -62$ because -37 is to the right of -62 on the number line.

▶ Absolute Value

The **absolute value** of a number is the distance the number is from zero on a number line. Absolute value is represented by the symbol $||$. Absolute values are *always* positive or zero.

Examples

$|-1| = 1$ The absolute value of -1 is 1. The distance of -1 from zero on a number line is 1.
$|1| = 1$ The absolute value of 1 is 1. The distance of 1 from zero on a number line is 1.
$|-23| = 23$ The absolute value of -23 is 23. The distance of -23 from zero on a number line is 23.
$|23| = 23$ The absolute value of 23 is 23. The distance of 23 from zero on a number line is 23.

The **absolute value** of an expression is the distance the value of the expression is from zero on a number line. Absolute values of expressions are *always* positive or zero.

Examples

$|3 - 5| = |-2| = 2$ The absolute value of $3 - 5$ is 2. The distance of $3 - 5$ from zero on a number line is 2.
$|5 - 3| = |2| = 2$ The absolute value of $5 - 3$ is 2. The distance of $5 - 3$ from zero on a number line is 2.

Practice Question

$|x - y| = 5$

Which values of x and y make the above equation NOT true?

a. $x = -8$ $y = -3$
b. $x = 12$ $y = 7$
c. $x = -20$ $y = -25$
d. $x = -5$ $y = 10$
e. $x = -2$ $y = 3$

Answer

d. Answer choice **a:** $|(-8) - (-3)| = |(-8) + 3| = |-5| = 5$
Answer choice **b:** $|12 - 7| = |5| = 5$
Answer choice **c:** $|(-20) - (-25)| = |(-20) + 25| = |5| = 5$
Answer choice **d:** $|(-5) - 10| = |-15| = 15$
Answer choice **e:** $|(-2) - 3| = |-5| = 5$
Therefore, the values of x and y in answer choice **d** make the equation NOT true.

▶ Rules for Working with Positive and Negative Integers

Multiplying/Dividing

- When multiplying or dividing two integers, if the signs are the same, the result is positive.

Examples

negative × positive = negative	$-3 \times 5 = -15$
positive ÷ positive = positive	$15 \div 5 = 3$
negative × negative = positive	$-3 \times -5 = 15$
negative ÷ negative = positive	$-15 \div -5 = 3$

- When multiplying or dividing two integers, if the signs are different, the result is negative:

Examples

positive × negative = negative	$3 \times -5 = -15$
positive ÷ negative = negative	$15 \div -5 = -3$

Adding

- When adding two integers with the same sign, the sum has the same sign as the addends.

Examples

positive + positive = positive	$4 + 3 = 7$
negative + negative = negative	$-4 + -3 = -7$

- When adding integers of different signs, follow this two-step process:

1. Subtract the absolute values of the numbers. Be sure to subtract the lesser absolute value from the greater absolute value.
2. Apply the sign of the larger number

Examples

$-2 + 6$

First subtract the absolute values of the numbers: $|6| - |-2| = 6 - 2 = 4$

Then apply the sign of the larger number: 6.

The answer is 4.

$7 + -12$

First subtract the absolute values of the numbers: $|-12| - |7| = 12 - 7 = 5$

Then apply the sign of the larger number: -12.

The answer is -5.

Subtracting

- When subtracting integers, change all subtraction to addition and change the sign of the number being subtracted to its opposite. Then follow the rules for addition.

Examples

$(+12) - (+15) = (+12) + (-15) = -3$

$(-6) - (-9) = (-6) + (+9) = +3$

Practice Question

Which of the following expressions is equal to -9?

a. $-17 + 12 - (-4) - (-10)$

b. $13 - (-7) - 36 - (-8)$

c. $-8 \times (-2) - 14 + (-11)$

d. $(-10 \times 4) - (-5 \times 5) - 6$

e. $[-48 \div (-3)] - (28 \div 4)$

Answer

c. Answer choice **a:** $-17 + 12 - (-4) - (-10) = 9$

Answer choice **b:** $13 - (-7) - 36 - (-8) = -8$

Answer choice **c:** $-8 \times (-2) - 14 + (-11) = -9$

Answer choice **d:** $(-10 \times 4) - (-5 \times 5) - 6 = -21$

Answer choice **e:** $[-48 \div (-3)] - (28 \div 4) = 9$

Therefore, answer choice **c** is equal to -9.

▶ Decimals

Memorize the order of place value:

3	7	5	9	•	1	6	0	4
THOUSANDS	HUNDREDS	TENS	ONES	DECIMAL POINT	TENTHS	HUNDREDTHS	THOUSANDTHS	TEN THOUSANDTHS

The number shown in the place value chart can also be expressed in expanded form:

3,759.1604 =

$(3 \times 1{,}000) + (7 \times 100) + (5 \times 10) + (9 \times 1) + (1 \times 0.1) + (6 \times 0.01) + (0 \times 0.001) + (4 \times 0.0001)$

Comparing Decimals

When comparing decimals less than one, line up the decimal points and fill in any zeroes needed to have an equal number of digits in each number.

Example

Compare 0.8 and 0.008.
Line up decimal points 0.8**00**
and add zeroes 0.008.
Then ignore the decimal point and ask, which is greater: 800 or 8?
800 is bigger than 8, so 0.8 is greater than 0.008.

Practice Question

Which of the following inequalities is true?

a. 0.04 < 0.004
b. 0.17 < 0.017
c. 0.83 < 0.80
d. 0.29 < 0.3
e. 0.5 < 0.08

Answer

d. Answer choice **a:** 0.0**40** > 0.**004** because 40 > 4. Therefore, 0.04 > 0.004. This answer choice is FALSE.
Answer choice **b:** 0.**170** > 0.**017** because 170 > 17. Therefore, 0.17 > 0.017. This answer choice is FALSE.
Answer choice **c:** 0.**83** > 0.**80** because 83 > 80. This answer choice is FALSE.
Answer choice **d:** 0.**29** < 0.**30** because 29 < 30. Therefore, 0.29 < 0.3. This answer choice is TRUE.
Answer choice **e:** 0.**50** > 0.**08** because 50 > 8. Therefore, 0.5 > 0.08. This answer choice is FALSE.

▶ Fractions

Multiplying Fractions

To multiply fractions, simply multiply the numerators and the denominators:

$\frac{a}{b} \times \frac{c}{d} = \frac{a \times c}{b \times d}$ $\frac{5}{8} \times \frac{3}{7} = \frac{5 \times 3}{8 \times 7} = \frac{15}{56}$ $\frac{3}{4} \times \frac{5}{6} = \frac{3 \times 5}{4 \times 6} = \frac{15}{24}$

Practice Question

Which of the following fractions is equivalent to $\frac{2}{9} \times \frac{3}{5}$?

a. $\frac{5}{45}$
b. $\frac{6}{45}$
c. $\frac{5}{14}$
d. $\frac{10}{18}$
e. $\frac{37}{45}$

Answer

b. $\frac{2}{9} \times \frac{3}{5} = \frac{2 \times 3}{9 \times 5} = \frac{6}{45}$

Reciprocals

To find the reciprocal of any fraction, swap its numerator and denominator.

Examples

Fraction: $\frac{1}{4}$ Reciprocal: $\frac{4}{1}$
Fraction: $\frac{5}{6}$ Reciprocal: $\frac{6}{5}$
Fraction: $\frac{7}{2}$ Reciprocal: $\frac{2}{7}$
Fraction: $\frac{x}{y}$ Reciprocal: $\frac{y}{x}$

Dividing Fractions

Dividing a fraction by another fraction is the same as multiplying the first fraction by the **reciprocal** of the second fraction:

$\frac{a}{b} \div \frac{c}{d} = \frac{a}{b} \times \frac{d}{c} = \frac{a \times d}{b \times c}$ $\frac{3}{4} \div \frac{2}{5} = \frac{3}{4} \times \frac{5}{2} = \frac{15}{8}$ $\frac{3}{4} \div \frac{5}{6} = \frac{3}{4} \times \frac{6}{5} = \frac{3 \times 6}{4 \times 5} = \frac{18}{20}$

Adding and Subtracting Fractions with Like Denominators

To add or subtract fractions with like denominators, add or subtract the numerators and leave the denominator as it is:

$\frac{a}{c} + \frac{b}{c} = \frac{a+b}{c}$ $\frac{1}{6} + \frac{4}{6} = \frac{1+4}{6} = \frac{5}{6}$

$\frac{a}{c} - \frac{b}{c} = \frac{a-b}{c}$ $\frac{5}{7} - \frac{3}{7} = \frac{5-3}{7} = \frac{2}{7}$

Adding and Subtracting Fractions with Unlike Denominators

To add or subtract fractions with unlike denominators, find the **Least Common Denominator**, or **LCD**, and convert the unlike denominators into the LCD. The LCD is the smallest number divisible by each of the denominators. For example, the LCD of $\frac{1}{8}$ and $\frac{1}{12}$ is 24 because 24 is the least multiple shared by 8 and 12. Once you know the LCD, convert each fraction to its new form by multiplying both the numerator and denominator by the necessary number to get the LCD, and then add or subtract the new numerators.

Example

$\frac{1}{8} + \frac{1}{12}$ LCD is 24 because $8 \times 3 = 24$ and $12 \times 2 = 24$.

$\frac{1}{8} = 1 \times \frac{3}{8} \times 3 = \frac{3}{24}$ Convert fraction.

$\frac{1}{12} = 1 \times \frac{2}{12} \times 2 = \frac{2}{24}$ Convert fraction.

$\frac{3}{24} + \frac{2}{24} = \frac{5}{24}$ Add numerators only.

Example

$\frac{4}{9} - \frac{1}{6}$ LCD is 54 because $9 \times 6 = 54$ and $6 \times 9 = 54$.

$\frac{4}{9} = 4 \times \frac{6}{9} \times 6 = \frac{24}{54}$ Convert fraction.

$\frac{1}{6} = 1 \times \frac{9}{6} \times 9 = \frac{9}{54}$ Convert fraction.

$\frac{24}{54} - \frac{9}{54} = \frac{15}{54} = \frac{5}{18}$ Subtract numerators only. Reduce where possible.

Practice Question

Which of the following expressions is equivalent to $\frac{5}{8} \div \frac{3}{4}$?

a. $\frac{1}{3} + \frac{1}{2}$
b. $\frac{3}{4} + \frac{5}{8}$
c. $\frac{1}{3} + \frac{2}{3}$
d. $\frac{4}{12} + \frac{1}{12}$
e. $\frac{1}{6} + \frac{3}{6}$

Answer

a. The expression in the equation is $\frac{5}{8} \div \frac{3}{4} = \frac{5}{8} \times \frac{4}{3} = \frac{5 \times 4}{8 \times 3} = \frac{20}{24} = \frac{5}{6}$. So you must evaluate each answer choice to determine which equals $\frac{5}{6}$.
Answer choice **a:** $\frac{1}{3} + \frac{1}{2} = \frac{2}{6} + \frac{3}{6} = \frac{5}{6}$.
Answer choice **b:** $\frac{3}{4} + \frac{5}{8} = \frac{6}{8} + \frac{5}{8} = \frac{11}{8}$.
Answer choice **c:** $\frac{1}{3} + \frac{2}{3} = \frac{3}{3} = \frac{6}{6} = 1$.
Answer choice **d:** $\frac{4}{12} + \frac{1}{12} = \frac{5}{12}$.
Answer choice **e:** $\frac{1}{6} + \frac{3}{6} = \frac{4}{6}$.
Therefore, answer choice **a** is correct.

▶ Sets

Sets are collections of certain numbers. All of the numbers within a set are called the **members** of the set.

Examples

The set of integers is $\{\ldots -3, -2, -1, 0, 1, 2, 3, \ldots\}$.

The set of whole numbers is $\{0, 1, 2, 3, \ldots\}$.

Intersections

When you find the elements that two (or more) sets have in common, you are finding the **intersection** of the sets. The symbol for intersection is $\cap$.

Example

The set of negative integers is $\{\ldots, -4, -3, -2, -1\}$.

The set of even numbers is $\{\ldots, -4, -2, 0, 2, 4, \ldots\}$.

The intersection of the set of negative integers and the set of even numbers is the set of elements (numbers) that the two sets have in common:

$\{\ldots, -8, -6, -4, -2\}$.

Practice Question

Set X = even numbers between 0 and 10

Set Y = prime numbers between 0 and 10

What is $X \cap Y$?

a. $\{1, 2, 3, 4, 5, 6, 7, 8, 9\}$

b. $\{1, 2, 3, 4, 5, 6, 7, 8\}$

c. $\{2\}$

d. $\{2, 4, 6, 8\}$

e. $\{1, 2, 3, 5, 7\}$

Answer

c. $X \cap Y$ is "the intersection of sets X and Y." The intersection of two sets is the set of numbers shared by both sets. Set $X = \{2, 4, 6, 8\}$. Set $Y = \{1, 2, 3, 5, 7\}$. Therefore, the intersection is $\{2\}$.

Unions

When you combine the elements of two (or more) sets, you are finding the **union** of the sets. The symbol for union is $\cup$.

Example

The positive even integers are $\{2, 4, 6, 8, \ldots\}$.

The positive odd integers are $\{1, 3, 5, 7, \ldots\}$.

If we combine the elements of these two sets, we find the union of these sets:

$\{1, 2, 3, 4, 5, 6, 7, 8, \ldots\}$.

Practice Question

Set $P = \{0, \frac{3}{7}, 0.93, 4, 6.98, \frac{27}{2}\}$

Set $Q = \{0.01, 0.15, 1.43, 4\}$

What is $P \cup Q$?

a. $\{4\}$

b. $\{\frac{3}{7}, \frac{27}{2}\}$

c. $\{0, 4\}$

d. $\{0, 0.01, 0.15, \frac{3}{7}, 0.93, 1.43, 6.98, \frac{27}{2}\}$

e. $\{0, 0.01, 0.15, \frac{3}{7}, 0.93, 1.43, 4, 6.98, \frac{27}{2}\}$

Answer

e. $P \cup Q$ is "the union of sets P and Q." The union of two sets is all the numbers from the two sets combined. Set $P = \{0, \frac{3}{7}, 0.93, 4, 6.98, \frac{27}{2}\}$. Set $Q = \{0.01, 0.15, 1.43, 4\}$. Therefore, the union is $\{0, 0.01, 0.15, \frac{3}{7}, 0.93, 1.43, 4, 6.98, \frac{27}{2}\}$.

Mean, Median, and Mode

To find the average, or **mean**, of a set of numbers, add all of the numbers together and divide by the quantity of numbers in the set.

$$\text{mean} = \frac{\text{sum of numbers in set}}{\text{quantity of numbers in set}}$$

Example

Find the mean of 9, 4, 7, 6, and 4.

$\frac{9+4+7+6+4}{5} = \frac{30}{5} = 6$ The denominator is 5 because there are five numbers in the set.

To find the **median** of a set of numbers, arrange the numbers in ascending order and find the middle value.

- If the set contains an odd number of elements, then simply choose the middle value.

Example

Find the median of the number set: 1, 5, 3, 7, 2.

First arrange the set in ascending order: 1, 2, 3, 5, 7.

Then choose the middle value: 3.

The median is 3.

- If the set contains an even number of elements, then average the two middle values.

Example

Find the median of the number set: 1, 5, 3, 7, 2, 8.

First arrange the set in ascending order: 1, 2, 3, 5, 7, 8.

Then choose the middle values: 3 and 5.

Find the average of the numbers 3 and 5: $\frac{3+5}{2} = \frac{8}{2} = 4$.

The median is 4.

The **mode** of a set of numbers is the number that occurs most frequently.

Example

For the number set 1, 2, 5, 3, 4, 2, 3, 6, 3, 7, the number 3 is the mode because it occurs three times. The other numbers occur only once or twice.

Practice Question

If the mode of a set of three numbers is 17, which of the following must be true?

I. The average is greater than 17.
II. The average is odd.
III. The median is 17.

a. none
b. I only
c. III only
d. I and III
e. I, II, and III

Answer

c. If the mode of a set of three numbers is 17, the set is $\{x, 17, 17\}$. Using that information, we can evaluate the three statements:

Statement I: *The average is greater than 17.*
If x is less than 17, then the average of the set will be less than 17. For example, if $x = 2$, then we can find the average:
$2 + 17 + 17 = 36$
$36 \div 3 = 12$
Therefore, the average would be 12, which is not greater than 17, so number I isn't necessarily true. Statement I is FALSE.

Statement II: *The average is odd.*
Because we don't know the third number of the set, we don't know that the average must be even. As we just learned, if the third number is 2, the average is 12, which is even, so statement II ISN'T NECESSARILY TRUE.

Statement III: *The median is 17.*
We know that the median is 17 because the median is the middle value of the three numbers in the set. If $X > 17$, the median is 17 because the numbers would be ordered: X, 17, 17. If $X < 17$, the median is still 17 because the numbers would be ordered: 17, 17, X. Statement III is TRUE.

Answer: Only statement III is NECESSARILY TRUE.

▶ Percent

A percent is a ratio that compares a number to 100. For example, 30% = $\frac{30}{100}$.

- To convert a decimal to a percentage, move the decimal point two units to the right and add a percentage symbol.
  0.65 = 65% 0.04 = 4% 0.3 = 30%
- One method of converting a fraction to a percentage is to first change the fraction to a decimal (by dividing the numerator by the denominator) and to then change the decimal to a percentage.
  $\frac{3}{5} = 0.60 = 60\%$ $\frac{1}{5} = 0.2 = 20\%$ $\frac{3}{8} = 0.375 = 37.5\%$
- Another method of converting a fraction to a percentage is to, if possible, convert the fraction so that it has a denominator of 100. The percentage is the new numerator followed by a percentage symbol.
  $\frac{3}{5} = \frac{60}{100} = 60\%$ $\frac{6}{25} = \frac{24}{100} = 24\%$
- To change a percentage to a decimal, move the decimal point two places to the left and eliminate the percentage symbol.
  64% = 0.64 87% = 0.87 7% = 0.07
- To change a percentage to a fraction, divide by 100 and reduce.
  $44\% = \frac{44}{100} = \frac{11}{25}$ $70\% = \frac{70}{100} = \frac{7}{10}$ $52\% = \frac{52}{100} = \frac{26}{50}$
- Keep in mind that any percentage that is 100 or greater converts to a number greater than 1, such as a whole number or a mixed number.
  500% = 5 275% = 2.75 or $2\frac{3}{4}$

Here are some conversions you should be familiar with:

FRACTION	DECIMAL	PERCENTAGE
$\frac{1}{2}$	0.5	50%
$\frac{1}{4}$	0.25	25%
$\frac{1}{3}$	0.333 . . .	$33.\overline{3}\%$
$\frac{2}{3}$	0.666 . . .	$66.\overline{6}\%$
$\frac{1}{10}$	0.1	10%
$\frac{1}{8}$	0.125	12.5%
$\frac{1}{6}$	0.1666 . . .	$16.\overline{6}\%$
$\frac{1}{5}$	0.2	20%

Practice Question

If $\frac{7}{25} < x < 0.38$, which of the following could be a value of x?

a. 20%
b. 26%
c. 34%
d. 39%
e. 41%

Answer

c. $\frac{7}{25} = \frac{28}{100} = 28\%$

$0.38 = 38\%$

Therefore, $28\% < x < 38\%$.

Only answer choice **c**, 34%, is greater than 28% and less than 38%.

▶ Graphs and Tables

The SAT includes questions that test your ability to analyze graphs and tables. Always read graphs and tables carefully before moving on to read the questions. Understanding the graph will help you process the information that is presented in the question. Pay special attention to headings and units of measure in graphs and tables.

Circle Graphs or Pie Charts

This type of graph is representative of a whole and is usually divided into percentages. Each section of the chart represents a portion of the whole. All the sections added together equal 100% of the whole.

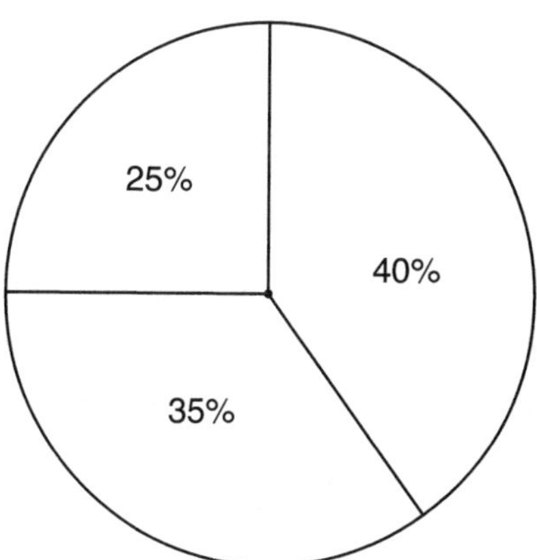

Bar Graphs

Bar graphs compare similar things with different length bars representing different values. On the SAT, these graphs frequently contain differently shaded bars used to represent different elements. Therefore, it is important to pay attention to both the size and shading of the bars.

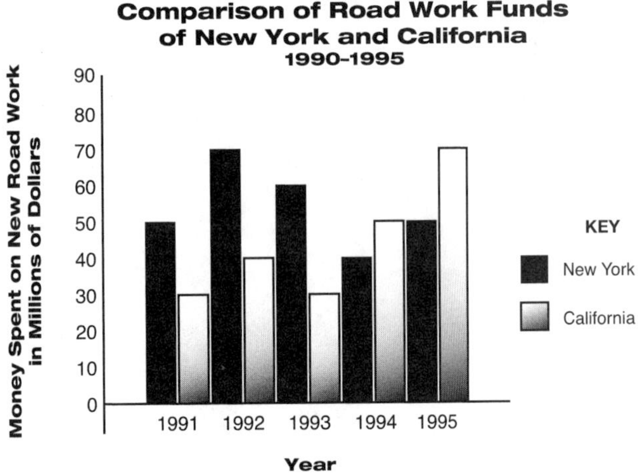

Broken-Line Graphs

Broken-line graphs illustrate a measurable change over time. If a line is slanted up, it represents an increase whereas a line sloping down represents a decrease. A flat line indicates no change as time elapses.

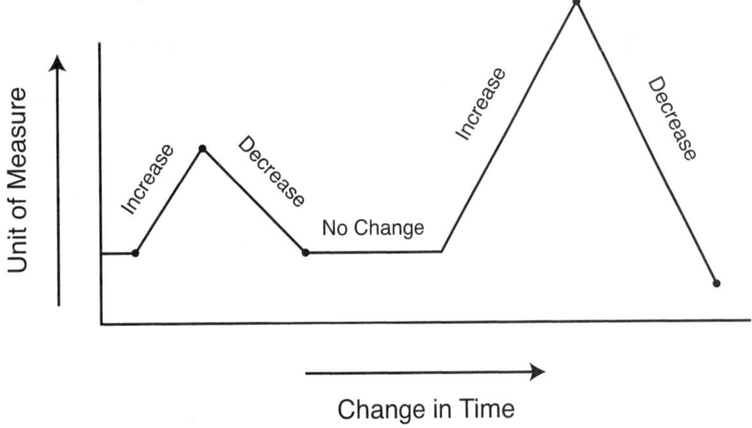

Scatterplots illustrate the relationship between two quantitative variables. Typically, the values of the independent variables are the x-coordinates, and the values of the dependent variables are the y-coordinates. When presented with a scatterplot, look for a trend. Is there a line that the points seem to cluster around? For example:

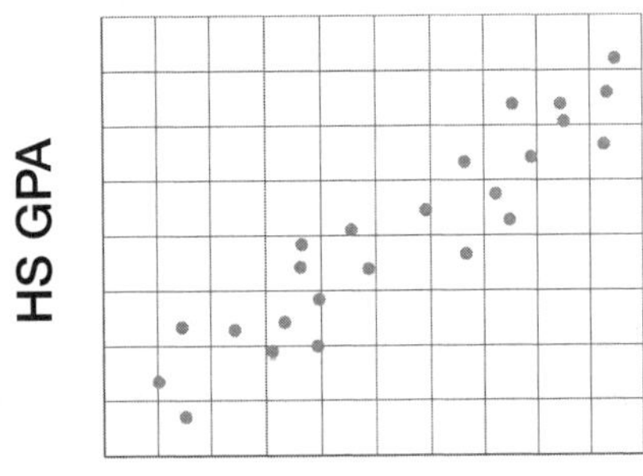

In the previous scatterplot, notice that a "line of best fit" can be created:

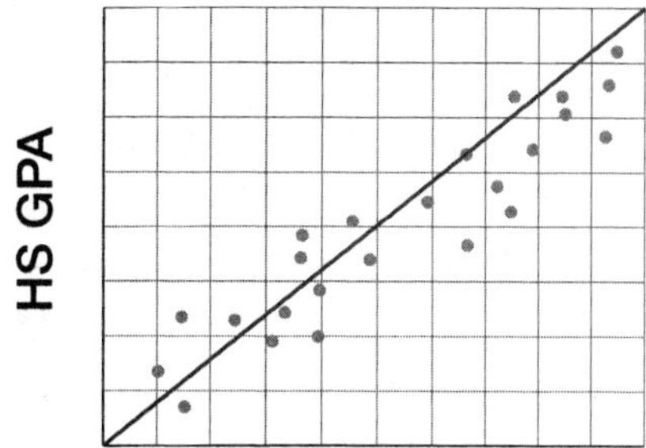

Practice Question

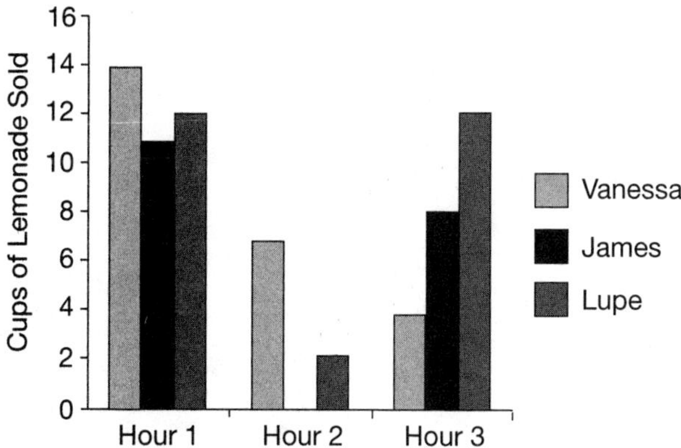

Based on the graph above, which of the following statements are true?

I. In the first hour, Vanessa sold the most lemonade.
II. In the second hour, Lupe didn't sell any lemonade.
III. In the third hour, James sold twice as much lemonade as Vanessa.

a. I only
b. II only
c. I and II
d. I and III
e. I, II, and III

Answer

d. Let's evaluate the three statements:

Statement I: *In the first hour, Vanessa sold the most lemonade.*

In the graph, Vanessa's bar for the first hour is highest, which means she sold the most lemonade in the first hour. Therefore, statement I is TRUE.

Statement II: *In the second hour, Lupe didn't sell any lemonade.*

In the second hour, there is no bar for James, which means he sold no lemonade. However, the bar for Lupe is at 2, so Lupe sold 2 cups of lemonade. Therefore, statement II is FALSE.

Statement III: *In the third hour, James sold twice as much lemonade as Vanessa.*
In the third hour, James's bar is at 8 and Vanessa's bar is at 4, which means James sold twice as much lemonade as Vanessa. Therefore, statement III is TRUE.

Answer: Only statements I and III are true.

▶ Matrices

Matrices are rectangular arrays of numbers. Below is an example of a 2 by 2 matrix:

$$\begin{bmatrix} a_1 & a_2 \\ a_3 & a_4 \end{bmatrix}$$

Review the following basic rules for performing operations on 2 by 2 matrices.

Addition

$$\begin{bmatrix} a_1 & a_2 \\ a_3 & a_4 \end{bmatrix} + \begin{bmatrix} b_1 & b_2 \\ b_3 & b_4 \end{bmatrix} = \begin{bmatrix} a_1 + b_1 & a_2 + b_2 \\ a_3 + b_3 & a_4 + b_4 \end{bmatrix}$$

Subtraction

$$\begin{bmatrix} a_1 & a_2 \\ a_3 & a_4 \end{bmatrix} - \begin{bmatrix} b_1 & b_2 \\ b_3 & b_4 \end{bmatrix} = \begin{bmatrix} a_1 - b_1 & a_2 - b_2 \\ a_3 - b_3 & a_4 - b_4 \end{bmatrix}$$

Multiplication

$$\begin{bmatrix} a_1 & a_2 \\ a_3 & a_4 \end{bmatrix} \times \begin{bmatrix} b_1 & b_2 \\ b_3 & b_4 \end{bmatrix} = \begin{bmatrix} a_1b_1 + a_2b_3 & a_1b_2 + a_2b_4 \\ a_3b_1 + a_4b_3 & a_3b_2 + a_4b_4 \end{bmatrix}$$

Scalar Multiplication

$$k\begin{bmatrix} a_1 & a_2 \\ a_3 & a_4 \end{bmatrix} = \begin{bmatrix} ka_1 & ka_2 \\ ka_3 & ka_4 \end{bmatrix}$$

Practice Question

$$\begin{bmatrix} 4 & 3 \\ 7 & 1 \end{bmatrix} + \begin{bmatrix} 6 & 2 \\ 5 & 2 \end{bmatrix} =$$

Which of the following shows the correct solution to the problem above?

a. $\begin{bmatrix} 7 & 8 \\ 8 & 7 \end{bmatrix}$

b. $\begin{bmatrix} 11 & 11 \\ 4 & 4 \end{bmatrix}$

c. $\begin{bmatrix} -2 & 1 \\ 2 & -1 \end{bmatrix}$

d. $\begin{bmatrix} 24 & 6 \\ 35 & 2 \end{bmatrix}$

e. $\begin{bmatrix} 10 & 5 \\ 12 & 3 \end{bmatrix}$

Answer

e. $\begin{bmatrix} 4 & 3 \\ 7 & 1 \end{bmatrix} + \begin{bmatrix} 6 & 2 \\ 5 & 2 \end{bmatrix} = \begin{bmatrix} 4+6 & 3+2 \\ 7+5 & 1+2 \end{bmatrix} = \begin{bmatrix} 10 & 5 \\ 12 & 3 \end{bmatrix}$

CHAPTER

Algebra Review

This chapter reviews key skills and concepts of algebra that you need to know for the SAT. Throughout the chapter are sample questions in the style of SAT questions. Each sample SAT question is followed by an explanation of the correct answer.

▶ Equations

To solve an algebraic **equation** with one variable, find the value of the unknown variable.

Rules for Working with Equations

1. The equal sign separates an equation into two sides.
2. Whenever an operation is performed on one side, the same operation must be performed on the other side.
3. To solve an equation, first move all of the variables to one side and all of the numbers to the other. Then simplify until only one variable (with a coefficient of 1) remains on one side and one number remains on the other side.

Example

$7x - 11 = 29 - 3x$	Move the variables to one side.
$7x - 11 + 3x = 29 - 3x + 3x$	Perform the same operation on both sides.
$10x - 11 = 29$	Now move the numbers to the other side.
$10x - 11 + 11 = 29 + 11$	Perform the same operation on both sides.
$10x = 40$	Divide both sides by the coefficient.
$\frac{10x}{10} = \frac{40}{10}$	Simplify.
$x = 4$	

Practice Question

If $13x - 28 = 22 - 12x$, what is the value of x?

a. -6
b. $-\frac{6}{25}$
c. 2
d. 6
e. 50

Answer

c. To solve for x:

$13x - 28 = 22 - 12x$
$13x - 28 + 12x = 22 - 12x + 12x$
$25x - 28 = 22$
$25x - 28 + 28 = 22 + 28$
$25x = 50$
$x = 2$

Cross Products

You can solve an equation that sets one fraction equal to another by finding **cross products** of the fractions. Finding cross products allows you to remove the denominators from each side of the equation by multiplying each side by a fraction equal to 1 that has the denominator from the opposite side.

Example

$\frac{a}{b} = \frac{c}{d}$	First multiply one side by $\frac{d}{d}$ and the other by $\frac{b}{b}$. The fractions $\frac{d}{d}$ and $\frac{b}{b}$ both equal 1, so they don't change the equation.
$\frac{a}{b} \times \frac{d}{d} = \frac{c}{d} \times \frac{b}{b}$	
$\frac{ad}{bd} = \frac{bc}{bd}$	The denominators are now the same. Now multiply both sides by the denominator and simplify.
$bd \times \frac{ad}{bd} = bd \times \frac{bc}{bd}$	
$ad = bc$	The example above demonstrates how finding cross products works. In the future, you can skip all the middle steps and just assume that $\frac{a}{b} = \frac{c}{d}$ is the same as $ad = bc$.

Example

$\frac{x}{6} = \frac{12}{36}$ Find cross products.

$36x = 6 \times 12$

$36x = 72$

$x = 2$

Example

$\frac{x}{4} = \frac{x + 12}{16}$ Find cross products.

$16x = 4(x + 12)$

$16x = 4x + 48$

$12x = 48$

$x = 4$

Practice Question

If $\frac{y}{9} = \frac{y-7}{12}$, what is the value of y?

a. -28
b. -21
c. $-\frac{63}{11}$
d. $-\frac{7}{3}$
e. 28

Answer

b. To solve for y:

$\frac{y}{9} = \frac{y-7}{12}$ Find cross products.

$12y = 9(y - 7)$

$12y = 9y - 63$

$12y - 9y = 9y - 63 - 9y$

$3y = -63$

$y = -21$

Checking Equations

After you solve an equation, you can check your answer by substituting your value for the variable into the original equation.

Example

We found that the solution for $7x - 11 = 29 - 3x$ is $x = 4$. To check that the solution is correct, substitute 4 for x in the equation:

$7x - 11 = 29 - 3x$

$7(4) - 11 = 29 - 3(4)$

$28 - 11 = 29 - 12$

$17 = 17$

This equation checks, so $x = 4$ is the correct solution!

Special Tips for Checking Equations on the SAT

1. If time permits, check all equations.
2. For questions that ask you to find the solution to an equation, you can simply substitute each answer choice into the equation and determine which value makes the equation correct. Begin with choice **c**. If choice **c** is not correct, pick an answer choice that is either larger or smaller.
3. Be careful to answer the question that is being asked. Sometimes, questions require that you solve for a variable and then perform an operation. For example, a question may ask the value of $x - 2$. You might find that $x = 2$ and look for an answer choice of 2. But the question asks for the value of $x - 2$ and the answer is not 2, but $2 - 2$. Thus, the answer is 0.

Equations with More Than One Variable

Some equations have more than one variable. To find the solution of these equations, solve for one variable in terms of the other(s). Follow the same method as when solving single-variable equations, but isolate only one variable.

Example

$3x + 6y = 24$ — To isolate the x variable, move $6y$ to the other side.

$3x + 6y - 6y = 24 - 6y$

$3x = 24 - 6y$

$\frac{3x}{3} = \frac{24 - 6y}{3}$ — Then divide both sides by 3, the coefficient of x.

$x = 8 - 2y$ — Then simplify. The solution is for x in terms of y.

Practice Question

If $8a + 16b = 32$, what does a equal in terms of b?

a. $4 - 2b$
b. $2 - \frac{1}{2}b$
c. $32 - 16b$
d. $4 - 16b$
e. $24 - 16b$

Answer

a. To solve for a in terms of b:

$8a + 16b = 32$

$8a + 16b - 16b = 32 - 16b$

$8a = 32 - 16b$

$\frac{8a}{8} = \frac{32 - 16b}{8}$

$a = 4 - 2b$

▶ Monomials

A **monomial** is an expression that is a number, a variable, or a product of a number and one or more variables.

6 y $-5xy^2$ $19a^6b^4$

▶ Polynomials

A **polynomial** is a monomial or the sum or difference of two or more monomials.

$7y^5$ $-6ab^4$ $8x + y^3$ $8x + 9y - z$

Operations with Polynomials

To add polynomials, simply combine like terms.

Example

$(5y^3 - 2y + 1) + (y^3 + 7y - 4)$

First remove the parentheses:

$5y^3 - 2y + 1 + y^3 + 7y - 4$

Then arrange the terms so that like terms are grouped together:

$5y^3 + y^3 - 2y + 7y + 1 - 4$

Now combine like terms:

Answer: $6y^3 + 5y - 3$

Example

$(2x - 5y + 8z) - (16x + 4y - 10z)$

First remove the parentheses. Be sure to distribute the subtraction correctly to all terms in the second set of parentheses:

$2x - 5y + 8z - 16x - 4y + 10z$

Then arrange the terms so that like terms are grouped together:

$2x - 16x - 5y - 4y + 8z + 10z$

Three Kinds of Polynomials

- A **monomial** is a polynomial with one term, such as $5b^6$.
- A **binomial** is a polynomial with two unlike terms, such as $2x + 4y$.
- A **trinomial** is a polynomial with three unlike terms, such as $y^3 + 8z - 2$.

Now combine like terms:

$-14x - 9y + 18z$

To multiply monomials, multiply their coefficients and multiply like variables by adding their exponents.

Example

$(-4a^3b)(6a^2b^3) = (-4)(6)(a^3)(a^2)(b)(b^3) = -24a^5b^4$

To divide monomials, divide their coefficients and divide like variables by subtracting their exponents.

Example

$\frac{10x^5y^7}{15x^4y^2} = (\frac{10}{15})(\frac{x^5}{x^4})(\frac{y^7}{y^2}) = \frac{2xy^5}{3}$

To multiply a polynomial by a monomial, multiply each term of the polynomial by the monomial and add the products.

Example

$8x(12x - 3y + 9)$	Distribute.
$(8x)(12x) - (8x)(3y) + (8x)(9)$	Simplify.
$96x^2 - 24xy + 72x$	

To divide a polynomial by a monomial, divide each term of the polynomial by the monomial and add the quotients.

Example

$\frac{6x - 18y + 42}{6} = \frac{6x}{6} - \frac{18y}{6} + \frac{42}{6} = x - 3y + 7$

Practice Question

Which of the following is the solution to $\frac{18x^8y^5}{24x^3y^4}$?

a. $\frac{3}{4x^5y}$

b. $\frac{18x^{11}y^9}{24}$

c. $42x^{11}y^9$

d. $\frac{3x^5y}{4}$

e. $\frac{x^5y}{6}$

Answer

d. To find the quotient:

$\frac{18x^8y^5}{24x^3y^4}$ Divide the coefficients and subtract the exponents.

$\frac{3x^{8-3}y^{5-4}}{4}$

$\frac{3x^5y^1}{4}$

$= \frac{3x^5y}{4}$

FOIL

The FOIL method is used when multiplying binomials. FOIL represents the order used to multiply the terms: **F**irst, **O**uter, **I**nner, and **L**ast. To multiply binomials, you multiply according to the FOIL order and then add the products.

Example

$(4x + 2)(9x + 8)$

F: $4x$ and $9x$ are the **first** pair of terms.

O: $4x$ and 8 are the **outer** pair of terms.

I: 2 and $9x$ are the **inner** pair of terms.

L: 2 and 8 are the **last** pair of terms.

Multiply according to FOIL:

$(4x)(9x) + (4x)(8) + (2)(9x) + (2)(8) = 36x^2 + 32x + 18x + 16$

Now combine like terms:

$36x^2 + 50x + 16$

Practice Question

Which of the following is the product of $7x + 3$ and $5x - 2$?

a. $12x^2 - 6x + 1$
b. $35x^2 + 29x - 6$
c. $35x^2 + x - 6$
d. $35x^2 + x + 6$
e. $35x^2 + 11x - 6$

Answer

c. To find the product, follow the FOIL method:

$(7x + 3)(5x - 2)$

F: $7x$ and $5x$ are the **first** pair of terms.

O: $7x$ and -2 are the **outer** pair of terms.

I: 3 and $5x$ are the **inner** pair of terms.

L: 3 and -2 are the **last** pair of terms.

Now multiply according to FOIL:

$(7x)(5x) + (7x)(-2) + (3)(5x) + (3)(-2) = 35x^2 - 14x + 15x - 6$

Now combine like terms:

$35x^2 + x - 6$

Factoring

Factoring is the reverse of multiplication. When multiplying, you find the product of factors. When factoring, you find the factors of a product.

Multiplication: $3(x + y) = 3x + 3y$
Factoring: $3x + 3y = 3(x + y)$

Three Basic Types of Factoring

- Factoring out a common monomial:

$18x^2 - 9x = 9x(2x - 1)$ $\quad ab - cb = b(a - c)$

- Factoring a quadratic trinomial using FOIL in reverse:

$x^2 - x - 20 = (x - 4)(x + 4)$ $\quad x^2 - 6x + 9 = (x - 3)(x - 3) = (x - 3)^2$

- Factoring the difference between two perfect squares using the rule $a^2 - b^2 = (a + b)(a - b)$:

$x^2 - 81 = (x + 9)(x - 9)$ $\quad x^2 - 49 = (x + 7)(x - 7)$

Practice Question

Which of the following expressions can be factored using the rule $a^2 - b^2 = (a + b)(a - b)$ where b is an integer?

a. $x^2 - 27$
b. $x^2 - 40$
c. $x^2 - 48$
d. $x^2 - 64$
e. $x^2 - 72$

Answer

d. The rule $a^2 - b^2 = (a + b)(a - b)$ applies to only the difference between perfect squares. 27, 40, 48, and 72 are not perfect squares. 64 is a perfect square, so $x^2 - 64$ can be factored as $(x + 8)(x - 8)$.

Using Common Factors

With some polynomials, you can determine a **common factor** for each term. For example, $4x$ is a common factor of all three terms in the polynomial $16x^4 + 8x^2 + 24x$ because it can divide evenly into each of them. To factor a polynomial with terms that have common factors, you can divide the polynomial by the known factor to determine the second factor.

Example

In the binomial $64x^3 + 24x$, $8x$ is the greatest common factor of both terms.

Therefore, you can divide $64x^3 + 24x$ by $8x$ to find the other factor.

$\frac{64x^3 + 24x}{8x} = \frac{64x^3}{8x} + \frac{24x}{8x} = 8x^2 + 3$

Thus, factoring $64x^3 + 24x$ results in $8x(8x^2 + 3)$.

Practice Question

Which of the following are the factors of $56a^5 + 21a$?

a. $7a(8a^4 + 3a)$
b. $7a(8a^4 + 3)$
c. $3a(18a^4 + 7)$
d. $21a(56a^4 + 1)$
e. $7a(8a^5 + 3a)$

Answer

b. To find the factors, determine a common factor for each term of $56a^5 + 21a$. Both coefficients (56 and 21) can be divided by 7 and both variables can be divided by a. Therefore, a common factor is $7a$. Now, to find the second factor, divide the polynomial by the first factor:

$\frac{56a^5 + 21a}{7a}$

$\frac{8a^5 + 3a}{a^1}$ Subtract exponents when dividing.

$8a^{5-1} + 3a^{1-1}$

$8a^4 + 3a^0$ A base with an exponent of $0 = 1$.

$8a^4 + 3(1)$

$8a^4 + 3$

Therefore, the factors of $56a^5 + 21a$ are $7a(8a^4 + 3)$.

Isolating Variables Using Fractions

It may be necessary to use factoring in order to isolate a variable in an equation.

Example

If $ax - c = bx + d$, what is x in terms of a, b, c, and d?

First isolate the x terms on the same side of the equation:

$ax - bx = c + d$

Now factor out the common x term:

$x(a - b) = c + d$

Then divide both sides by $a - b$ to isolate the variable x:

$\frac{x(a-b)}{a-b} = \frac{c+d}{a-b}$

Simplify:

$x = \frac{c+d}{a-b}$

Practice Question

If $bx + 3c = 6a - dx$, what does x equal in terms of a, b, c, and d?

a. $b - d$

b. $6a - 5c - b - d$

c. $(6a - 5c)(b + d)$

d. $\frac{6a - d - 5c}{b}$

e. $\frac{6a - 5c}{b + d}$

Answer

e. Use factoring to isolate x:

$bx + 5c = 6a - dx$	First isolate the x terms on the same side.
$bx + 5c + dx = 6a - dx + dx$	
$bx + 5c + dx = 6a$	
$bx + 5c + dx - 5c = 6a - 5c$	Finish isolating the x terms on the same side.
$bx + dx = 6a - 5c$	Now factor out the common x term.
$x(b + d) = 6a - 5c$	Now divide to isolate x.
$\frac{x(b + d)}{b + d} = \frac{6a - 5c}{b + d}$	
$x = \frac{6a - 5c}{b + d}$	

▶ Quadratic Trinomials

A **quadratic trinomial** contains an x^2 term as well as an x term. For example, $x^2 - 6x + 8$ is a quadratic trinomial. You can factor quadratic trinomials by using the FOIL method in reverse.

Example

Let's factor $x^2 - 6x + 8$.

Start by looking at the last term in the trinomial: 8. Ask yourself, "What two integers, when multiplied together, have a product of positive 8?" Make a mental list of these integers:

1×8 $\quad -1 \times -8$ $\quad 2 \times 4$ $\quad -2 \times -4$

Next look at the middle term of the trinomial: $-6x$. Choose the two factors from the above list that also add up to the coefficient -6:

-2 and -4

Now write the factors using -2 and -4:

$(x - 2)(x - 4)$

Use the FOIL method to double-check your answer:

$(x - 2)(x - 4) = x^2 - 6x + 8$

The answer is correct.

Practice Question

Which of the following are the factors of $z^2 - 6z + 9$?

a. $(z + 3)(z + 3)$
b. $(z + 1)(z + 9)$
c. $(z - 1)(z - 9)$
d. $(z - 3)(z - 3)$
e. $(z + 6)(z + 3)$

Answer

d. To find the factors, follow the FOIL method in reverse:

$z^2 - 6z + 9$

The product of the **last** pair of terms equals $+9$. There are a few possibilities for these terms: 3 and 3 (because $3 \times 3 = +9$), -3 and -3 (because $-3 \times -3 = +9$), 9 and 1 (because $9 \times 1 = +9$), -9 and -1 (because $-9 \times -1 = +9$).

The sum of the product of the **outer** pair of terms and the **inner** pair of terms equals $-6z$. So we must choose the two last terms from the list of possibilities that would add up to -6. The only possibility is -3 and -3. Therefore, we know the last terms are -3 and -3.

The product of the **first** pair of terms equals z^2. The most likely two terms for the first pair is z and z because $z \times z = z^2$.

Therefore, the factors are $(z - 3)(z - 3)$.

Fractions with Variables

You can work with fractions with variables the same as you would work with fractions without variables.

Example

Write $\frac{x}{6} - \frac{x}{12}$ as a single fraction.

First determine the LCD of 6 and 12: The LCD is 12. Then convert each fraction into an equivalent fraction with 12 as the denominator:

$\frac{x}{6} - \frac{x}{12} = \frac{x \times 2}{6 \times 2} - \frac{x}{12} = \frac{2x}{12} - \frac{x}{12}$

Then simplify:

$\frac{2x}{12} - \frac{x}{12} = \frac{x}{12}$

Practice Question

Which of the following best simplifies $\frac{5x}{8} - \frac{2x}{5}$?

a. $\frac{9}{40}$
b. $\frac{9x}{40}$
c. $\frac{x}{5}$
d. $\frac{3x}{40}$
e. x

Answer

b. To simplify the expression, first determine the LCD of 8 and 5: The LCD is 40. Then convert each fraction into an equivalent fraction with 40 as the denominator:

$\frac{5x}{8} - \frac{2x}{5} = (5x \times \frac{5}{8} \times 5) - \frac{(2x \times 8)}{(5 \times 8)} = \frac{25x}{40} - \frac{16x}{40}$

Then simplify:

$\frac{25x}{40} - \frac{16x}{40} = \frac{9x}{40}$

Reciprocal Rules

There are special rules for the sum and difference of reciprocals. The following formulas can be memorized for the SAT to save time when answering questions about reciprocals:

- If x and y are not 0, then $\frac{1}{x} + y = \frac{x+y}{xy}$
- If x and y are not 0, then $\frac{1}{x} - \frac{1}{y} = \frac{y-x}{xy}$

Note: These rules are easy to figure out using the techniques of the last section, if you are comfortable with them and don't like having too many formulas to memorize.

Quadratic Equations

A **quadratic equation** is an equation in the form $ax^2 + bx + c = 0$, where a, b, and c are numbers and $a \neq 0$. For example, $x^2 + 6x + 10 = 0$ and $6x^2 + 8x - 22 = 0$ are quadratic equations.

Zero-Product Rule

Because quadratic equations can be written as an expression equal to zero, the zero-product rule is useful when solving these equations.

The **zero-product rule** states that if the product of two or more numbers is 0, then at least one of the numbers is 0. In other words, if $ab = 0$, then you know that either a or b equals zero (or they both might be zero). This idea also applies when a and b are factors of an equation. When an equation equals 0, you know that one of the factors of the equation must equal zero, so you can determine the two possible values of x that make the factors equal to zero.

Example

Find the two possible values of x that make this equation true: $(x + 4)(x - 2) = 0$

Using the zero-product rule, you know that either $x + 4 = 0$ or that $x - 2 = 0$.

So solve both of these possible equations:

$x + 4 = 0$	$x - 2 = 0$
$x + 4 - 4 = 0 - 4$	$x - 2 + 2 = 0 + 2$
$x = -4$	$x = 2$

Thus, you know that both $x = -4$ and $x = 2$ will make $(x + 4)(x - 2) = 0$.

The zero product rule is useful when solving quadratic equations because you can rewrite a quadratic equation as equal to zero and take advantage of the fact that one of the factors of the quadratic equation is thus equal to 0.

Practice Question

If $(x - 8)(x + 5) = 0$, what are the two possible values of x?

a. $x = 8$ and $x = -5$
b. $x = -8$ and $x = 5$
c. $x = 8$ and $x = 0$
d. $x = 0$ and $x = -5$
e. $x = 13$ and $x = -13$

Answer

a. If $(x - 8)(x + 5) = 0$, then one (or both) of the factors must equal 0.
$x - 8 = 0$ if $x = 8$ because $8 - 8 = 0$.
$x + 5 = 0$ if $x = -5$ because $-5 + 5 = 0$.

Therefore, the two values of x that make $(x - 8)(x + 5) = 0$ are $x = 8$ and $x = -5$.

Solving Quadratic Equations by Factoring

If a quadratic equation is not equal to zero, rewrite it so that you can solve it using the zero-product rule.

Example

If you need to solve $x^2 - 11x = 12$, subtract 12 from both sides:
$x^2 - 11x - 12 = 12 - 12$
$x^2 - 11x - 12 = 0$
Now this quadratic equation can be solved using the zero-product rule:
$x^2 - 11x - 12 = 0$
$(x - 12)(x + 1) = 0$
Therefore:

$x - 12 = 0$	or	$x + 1 = 0$
$x - 12 + 12 = 0 + 12$		$x + 1 - 1 = 0 - 1$
$x = 12$		$x = -1$

Thus, you know that both $x = 12$ and $x = -1$ will make $x^2 - 11x - 12 = 0$.

A quadratic equation must be factored before using the zero-product rule to solve it.

Example

To solve $x^2 + 9x = 0$, first factor it:
$x(x + 9) = 0$.
Now you can solve it.
Either $x = 0$ or $x + 9 = 0$.
Therefore, possible solutions are $x = 0$ and $x = -9$.

Practice Question

If $x^2 - 8x = 20$, which of the following could be a value of $x^2 + 8x$?

a. -20
b. 20
c. 28
d. 108
e. 180

Answer

e. This question requires several steps to answer. First, you must determine the possible values of x considering that $x^2 - 8x = 20$. To find the possible x values, rewrite $x^2 - 8x = 20$ as $x^2 - 8x - 20 = 0$, factor, and then use the zero-product rule.

$x^2 - 8x - 20 = 0$ is factored as $(x - 10)(x + 2)$.

Thus, possible values of x are $x = 10$ and $x = -2$ because $10 - 10 = 0$ and $-2 + 2 = 0$.

Now, to find possible values of $x^2 + 8x$, plug in the x values:

If $x = -2$, then $x^2 + 8x = (-2)^2 + (8)(-2) = 4 + (-16) = -12$. None of the answer choices is -12, so try $x = 10$.

If $x = 10$, then $x^2 + 8x = 10^2 + (8)(10) = 100 + 80 = 180$.

Therefore, answer choice **e** is correct.

▶ Graphs of Quadratic Equations

The (x,y) solutions to quadratic equations can be plotted on a graph. It is important to be able to look at an equation and understand what its graph will look like. You must be able to determine what calculation to perform on each x value to produce its corresponding y value.

For example, below is the graph of $y = x^2$.

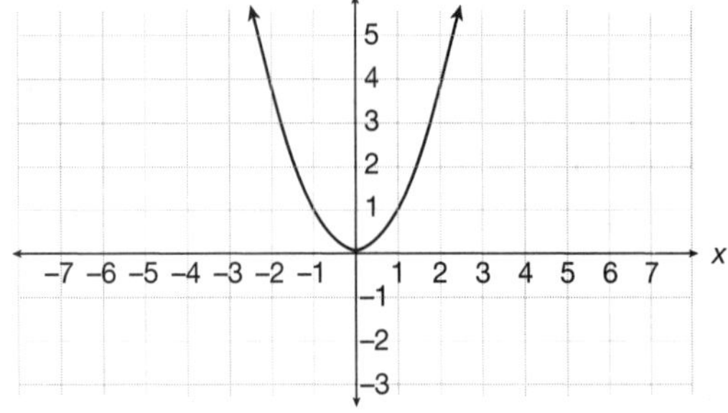

The equation $y = x^2$ tells you that for every x value, you must square the x value to find its corresponding y value. Let's explore the graph with a few x-coordinates:

An x value of 1 produces what y value? Plug $x = 1$ into $y = x^2$.

When $x = 1$, $y = 1^2$, so $y = 1$.
Therefore, you know a coordinate in the graph of $y = x^2$ is (1,1).

An x value of 2 produces what y value? Plug $x = 2$ into $y = x^2$.
When $x = 2$, $y = 2^2$, so $y = 4$.
Therefore, you know a coordinate in the graph of $y = x^2$ is (2,4).

An x value of 3 produces what y value? Plug $x = 3$ into $y = x^2$.
When $x = 3$, $y = 3^2$, so $y = 9$.
Therefore, you know a coordinate in the graph of $y = x^2$ is (3,9).

The SAT may ask you, for example, to compare the graph of $y = x^2$ with the graph of $y = (x - 1)^2$. Let's compare what happens when you plug numbers (x values) into $y = (x - 1)^2$ with what happens when you plug numbers (x values) into $y = x^2$:

$y = x^2$	$y = (x - 1)^2$
If $x = 1$, $y = 1$.	If $x = 1$, $y = 0$.
If $x = 2$, $y = 4$.	If $x = 2$, $y = 1$.
If $x = 3$, $y = 9$.	If $x = 3$, $y = 4$.
If $x = 4$, $y = 16$.	If $x = 4$, $y = 9$.

The two equations have the same y values, but they match up with different x values because $y = (x - 1)^2$ subtracts 1 before squaring the x value. As a result, the graph of $y = (x - 1)^2$ looks identical to the graph of $y = x^2$ except that the base is shifted to the right (on the x-axis) by 1:

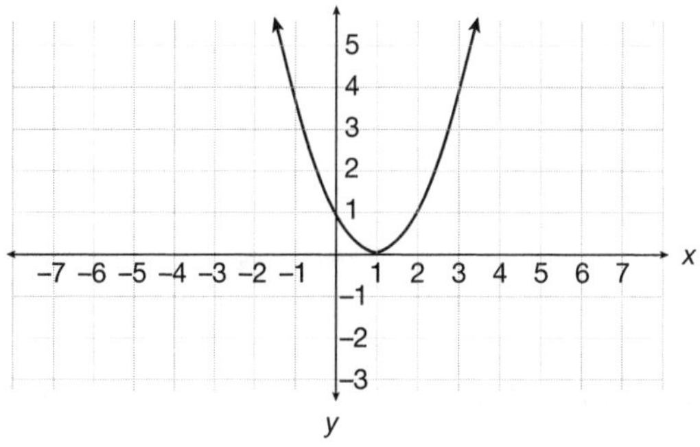

How would the graph of $y = x^2$ compare with the graph of $y = x^2 - 1$?

In order to find a y value with $y = x^2$, you square the x value. In order to find a y value with $y = x^2 - 1$, you square the x value and then subtract 1. This means the graph of $y = x^2 - 1$ looks identical to the graph of $y = x^2$ except that the base is shifted down (on the y-axis) by 1:

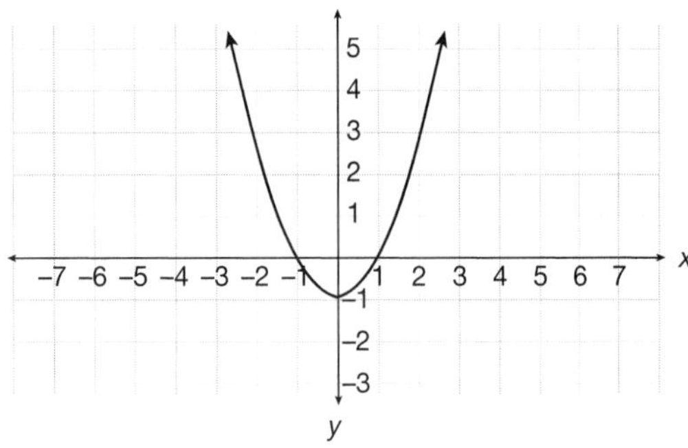

Practice Question

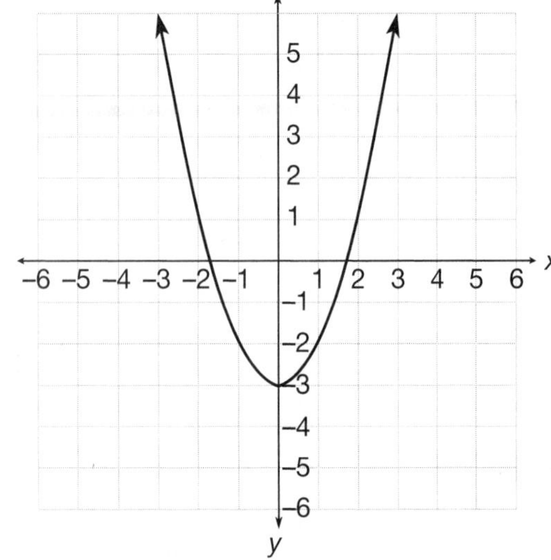

What is the equation represented in the graph above?

a. $y = x^2 + 3$
b. $y = x^2 - 3$
c. $y = (x + 3)^2$
d. $y = (x - 3)^2$
e. $y = (x - 1)^3$

Answer

b. This graph is identical to a graph of $y = x^2$ except it is moved down 3 so that the parabola intersects the y-axis at -3 instead of 0. Each y value is 3 less than the corresponding y value in $y = x^2$, so its equation is therefore $y = x^2 - 3$.

▶ Rational Equations and Inequalities

Rational numbers are numbers that can be written as fractions (and decimals and repeating decimals). Similarly, **rational equations** are equations in fraction form. **Rational inequalities** are also in fraction form and use the symbols $<$, $>$, $\leq$, and $\geq$ instead of $=$.

Example

Given $\frac{(x+5)(x^2+5x-14)}{(x^2+3x-10)} = 30$, find the value of x.

Factor the top and bottom:

$\frac{(x+5)(x+7)(x-2)}{(x+5)(x-2)} = 30$

You can cancel out the $(x + 5)$ and the $(x - 2)$ terms from the top and bottom to yield:

$x + 7 = 30$

Now solve for x:

$x + 7 = 30$

$x + 7 - 7 = 30 - 7$

$x = 23$

Practice Question

If $\frac{(x+8)(x^2+11x-26)}{(x^2+6x-16)} = 17$, what is the value of x?

a. -16
b. -13
c. -8
d. 2
e. 4

Answer

e. To solve for x, first factor the top and bottom of the fractions:

$\frac{(x+8)(x^2+11x-26)}{(x^2+6x-16)} = 17$

$\frac{(x+8)(x+13)(x-2)}{(x+8)(x-2)} = 17$

You can cancel out the $(x + 8)$ and the $(x - 2)$ terms from the top and bottom:

$x + 13 = 17$

Solve for x:

$x + 13 - 13 = 17 - 13$

$x = 4$

▶ Radical Equations

Some algebraic equations on the SAT include the square root of the unknown. To solve these equations, first isolate the radical. Then square both sides of the equation to remove the radical sign.

Example

$5\sqrt{c} + 15 = 35$

To isolate the variable, subtract 15 from both sides:

$5\sqrt{c} + 15 - 15 = 35 - 15$

$5\sqrt{c} = 20$

Next, divide both sides by 5:

$\frac{5\sqrt{c}}{5} = \frac{20}{5}$

$\sqrt{c} = 4$

Last, square both sides:

$(\sqrt{c})^2 = 4^2$

$c = 16$

Practice Question

If $6\sqrt{d} - 10 = 32$, what is the value of d?

a. 7
b. 14
c. 36
d. 49
e. 64

Answer

d. To solve for d, isolate the variable:

$6\sqrt{d} - 10 = 32$

$6\sqrt{d} - 10 + 10 = 32 + 10$

$6\sqrt{d} = 42$

$\frac{6\sqrt{d}}{6} = \frac{42}{6}$

$\sqrt{d} = 7$

$(\sqrt{d})^2 = 7^2$

$d = 49$

▶ Sequences Involving Exponential Growth

When analyzing a sequence, try to find the mathematical operation that you can perform to get the next number in the sequence. Let's try an example. Look carefully at the following sequence:

2, 4, 8, 16, 32, . . .

Notice that each successive term is found by multiplying the prior term by 2. ($2 \times 2 = 4$, $4 \times 2 = 8$, and so on.) Since each term is multiplied by a constant number (2), there is a constant ratio between the terms. Sequences that have a constant ratio between terms are called **geometric sequences.**

On the SAT, you may be asked to determine a specific term in a sequence. For example, you may be asked to find the thirtieth term of a geometric sequence like the previous one. You could answer such a question by writing out 30 terms of a sequence, but this is an inefficient method. It takes too much time. Instead, there is a formula to use. Let's determine the formula:

First, let's evaluate the terms.
2, 4, 8, 16, 32, . . .
Term 1 = 2
Term 2 = 4, which is 2×2
Term 3 = 8, which is $2 \times 2 \times 2$
Term 4 = 16, which is $2 \times 2 \times 2 \times 2$

You can also write out each term using exponents:

Term 1 = 2
Term 2 = 2×2^1
Term 3 = 2×2^2
Term 4 = 2×2^3

We can now write a formula:

Term $n = 2 \times 2^{n-1}$

So, if the SAT asks you for the thirtieth term, you know that:

Term $30 = 2 \times 2^{30-1} = 2 \times 2^{29}$

The generic formula for a geometric sequence is Term $n = a_1 \times r^{n-1}$, where n is the term you are looking for, a_1 is the first term in the series, and r is the ratio that the sequence increases by. In the above example, $n = 30$ (the thirtieth term), $a_1 = 2$ (because 2 is the first term in the sequence), and $r = 2$ (because the sequence increases by a ratio of 2; each term is two times the previous term).

You can use the formula Term $n = a_1 \times r^{n-1}$ when determining a term in any geometric sequence.

Practice Question

1, 3, 9, 27, 81, . . .

What is the thirty-eighth term of the sequence above?

a. 3^{38}

b. 3×1^{37}

c. 3×1^{38}

d. 1×3^{37}

e. 1×3^{38}

Answer

d. 1, 3, 9, 27, 81, . . . is a geometric sequence. There is a constant ratio between terms. Each term is three times the previous term. You can use the formula Term $n = a_1 \times r^{n-1}$ to determine the nth term of this geometric sequence.

First determine the values of n, a_1, and r:

$n = 38$ (because you are looking for the thirty-eighth term)

$a_1 = 1$ (because the first number in the sequence is 1)

$r = 3$ (because the sequence increases by a ratio of 3; each term is three times the previous term.)

Now solve:

Term $n = a_1 \times r^{n-1}$

Term $38 = 1 \times 3^{38-1}$

Term $38 = 1 \times 3^{37}$

▶ Systems of Equations

A system of equations is a set of two or more equations with the same solution. If $2c + d = 11$ and $c + 2d = 13$ are presented as a system of equations, we know that we are looking for values of c and d, which will be the same in both equations and will make both equations true.

Two methods for solving a system of equations are **substitution** and **linear combination.**

Substitution

Substitution involves solving for one variable in terms of another and then substituting that expression into the second equation.

Example

Here are the two equations with the same solution mentioned above:

$2c + d = 11$ and $c + 2d = 13$

To solve, first choose one of the equations and rewrite it, isolating one variable in terms of the other. It does not matter which variable you choose.

$2c + d = 11$ becomes $d = 11 - 2c$

Next substitute $11 - 2c$ for d in the other equation and solve:

$c + 2d = 13$
$c + 2(11 - 2c) = 13$
$c + 22 - 4c = 13$
$22 - 3c = 13$
$22 = 13 + 3c$
$9 = 3c$
$c = 3$
Now substitute this answer into either original equation for c to find d.
$2c + d = 11$
$2(3) + d = 11$
$6 + d = 11$
$d = 5$
Thus, $c = 3$ and $d = 5$.

Linear Combination

Linear combination involves writing one equation over another and then adding or subtracting the like terms so that one letter is eliminated.

Example
$x - 7 = 3y$ and $x + 5 = 6y$
First rewrite each equation in the same form.
$x - 7 = 3y$ becomes $x - 3y = 7$
$x + 5 = 6y$ becomes $x - 6y = -5$.
Now subtract the two equations so that the x terms are eliminated, leaving only one variable:

$$\begin{array}{r} x - 3y = 7 \\ \underline{-\ (x - 6y = -5)} \\ (x - x) + (-3y + 6y) = 7 - (-5) \\ 3y = 12 \\ y = 4 \text{ is the answer.} \end{array}$$

Now substitute 4 for y in one of the original equations and solve for x.
$x - 7 = 3y$
$x - 7 = 3(4)$
$x - 7 = 12$
$x - 7 + 7 = 12 + 7$
$x = 19$
Therefore, the solution to the system of equations is $y = 4$ and $x = 19$.

Systems of Equations with No Solution

It is possible for a system of equations to have no solution if there are no values for the variables that would make all the equations true. For example, the following system of equations has no solution because there are no values of x and y that would make both equations true:

$3x + 6y = 14$
$3x + 6y = 9$
In other words, one expression cannot equal both 14 and 9.

Practice Question

$5x + 3y = 4$
$15x + dy = 21$
What value of d would give the system of equations NO solution?

a. -9
b. -3
c. 1
d. 3
e. 9

Answer

e. The first step in evaluating a system of equations is to write the equations so that the coefficients of one of the variables are the same. If we multiply $5x + 3y = 4$ by 3, we get $15x + 9y = 12$. Now we can compare the two equations because the coefficients of the x variables are the same:
$15x + 9y = 12$
$15x + dy = 21$
The only reason there would be no solution to this system of equations is if the system contains the same expressions equaling different numbers. Therefore, we must choose the value of d that would make $15x + dy$ identical to $15x + 9y$. If $d = 9$, then:
$15x + 9y = 12$
$15x + 9y = 21$
Thus, if $d = 9$, there is no solution. Answer choice **e** is correct.

► Functions, Domain, and Range

A **function** is a relationship in which one value depends upon another value. Functions are written in the form beginning with the following symbols:

$f(x) =$

For example, consider the function $f(x) = 8x - 2$. If you are asked to find $f(3)$, you simply substitute the 3 into the given function equation.

$f(x) = 8x - 2$
becomes
$f(3) = 8(3) - 2f(3) = 24 - 2 = 22$

So, when $x = 3$, the value of the function is 22.

Potential functions must pass the **vertical line test** in order to be considered a function. The vertical line test is the following: Does any vertical line drawn through a graph of the potential function pass through only one point of the graph? If YES, then any vertical line drawn passes through only one point, and the potential function is a function. If NO, then a vertical line can be drawn that passes through more than one point, and the potential function is *not* a function.

The graph below shows a function because any vertical line drawn on the graph (such as the dotted vertical line shown) passes through the graph of the function only once:

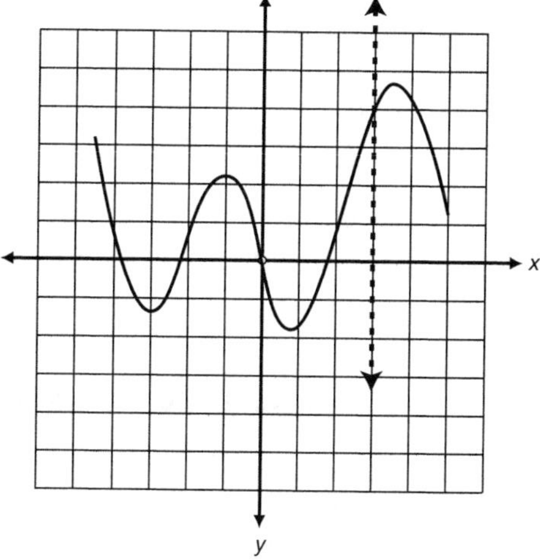

The graph below does NOT show a function because the dotted vertical line passes five times through the graph:

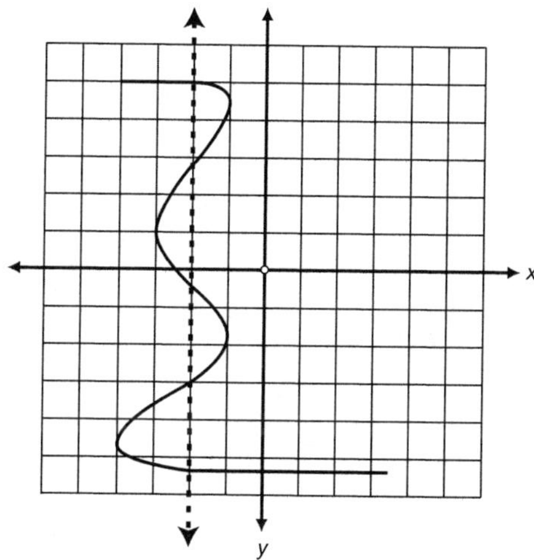

All of the x values of a function, collectively, are called its **domain**. Sometimes there are x values that are outside of the domain, but these are the x values for which the function is not defined.

All of the values taken on by $f(x)$ are collectively called the **range**. Any values that $f(x)$ cannot be equal to are said to be outside of the range.

The x values are known as the **independent variables**. The y values *depend* on the x values, so the y values are called the **dependent variables.**

Practice Question

If the function f is defined by $f(x) = 9x + 3$, which of the following is equal to $f(4b)$?

a. $36b + 12b$

b. $36b + 12$

c. $36b + 3$

d. $\frac{9}{4b+3}$

e. $\frac{4b}{9+3}$

Answer

c. If $f(x) = 9x + 3$, then, for $f(4b)$, $4b$ simply replaces x in $9x + 3$. Therefore, $f(4b) = 9(4b) + 3 = 36b + 3$.

Qualitative Behavior of Graphs and Functions

For the SAT, you should be able to analyze the graph of a function and interpret, qualitatively, something about the function itself.

Example

Consider the portion of the graph shown below. Let's determine how many values there are for $f(x) = 2$.

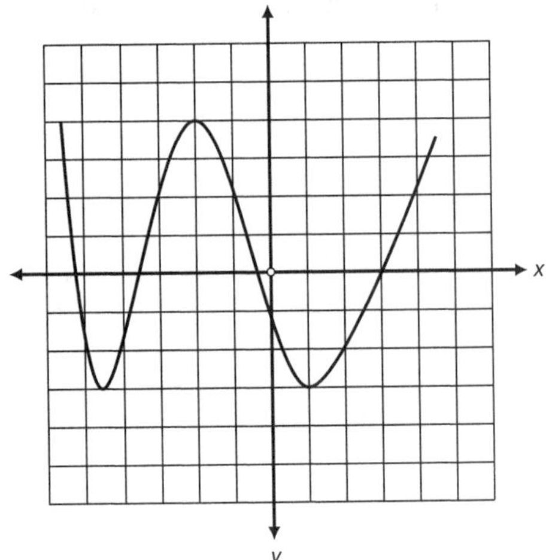

When $f(x) = 2$, the y value equals 2. So let's draw a horizontal line through $y = 2$ to see how many times the line intersects with the function. These points of intersection tell us the x values for $f(x) = 2$. As shown below, there are 4 such points, so we know there are four values for $f(x) = 2$.

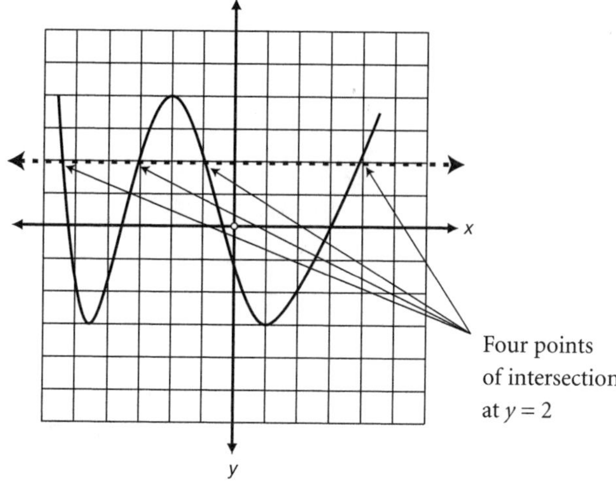

CHAPTER

7 Geometry Review

This chapter reviews key skills and concepts of geometry that you need to know for the SAT. Throughout the chapter are sample questions in the style of SAT questions. Each sample SAT question is followed by an explanation of the correct answer.

▶ Vocabulary

It is essential in geometry to recognize and understand the terminology used. Before you take the SAT, be sure you know and understand each geometry term in the following list.

acute angle	an angle that measures less than 90°
acute triangle	a triangle with every angle that measures less than 90°
adjacent angles	two angles that have the same vertex, share one side, and do not overlap
angle	two rays connected by a vertex
arc	a curved section of a circle
area	the number of square units inside a shape
bisect	divide into two equal parts
central angle	an angle formed by an arc in a circle

chord	a line segment that goes through a circle, with its endpoints on the circle
circumference	the distance around a circle
complementary angles	two angles whose sum is 90°
congruent	identical in shape and size; the geometric symbol for *congruent to* is ≅
coordinate plane	a grid divided into four quadrants by both a horizontal *x*-axis and a vertical *y*-axis
coordinate points	points located on a coordinate plane
diagonal	a line segment between two non-adjacent vertices of a polygon
diameter	a chord that passes through the center of a circle—the longest line you can draw in a circle. The term is used not only for this line segment, but also for its length.
equiangular polygon	a polygon with all angles of equal measure
equidistant	the same distance
equilateral triangle	a triangle with three equal sides and three equal angles
exterior angle	an angle on the outer sides of two lines cut by a transversal; or, an angle outside a triangle
hypotenuse	the longest leg of a right triangle. The hypotenuse is always opposite the right angle in a right triangle.
interior angle	an angle on the inner sides of two lines cut by a transversal
isosceles triangle	a triangle with two equal sides
line	a straight path that continues infinitely in two directions. The geometric notation for a line through points *A* and *B* is $\overline{AB}$.
line segment	the part of a line between (and including) two points. The geometric notation for the line segment joining points *A* and *B* is $\overline{AB}$. The notation $\overline{AB}$ is used both to refer to the segment itself and to its length.
major arc	an arc greater than or equal to 180°
midpoint	the point at the exact middle of a line segment
minor arc	an arc less than or equal to 180°
obtuse angle	an angle that measures greater than 90°
obtuse triangle	a triangle with an angle that measures greater than 90°
ordered pair	a location of a point on the coordinate plane in the form of (*x*,*y*). The *x* represents the location of the point on the horizontal *x*-axis, and the *y* represents the location of the point on the vertical *y*-axis.

origin	coordinate point (0,0): the point on a coordinate plane at which the *x*-axis and *y*-axis intersect
parallel lines	two lines in a plane that do not intersect. Parallel lines are marked by a symbol $\parallel$.
parallelogram	a quadrilateral with two pairs of parallel sides
perimeter	the distance around a figure
perpendicular lines	lines that intersect to form right angles
polygon	a closed figure with three or more sides
Pythagorean theorem	the formula $a^2 + b^2 = c^2$, where *a* and *b* represent the lengths of the *legs* and *c* represents the length of the *hypotenuse* of a right triangle
Pythagorean triple	a set of three whole numbers that satisfies the Pythagorean theorem, $a^2 + b^2 = c^2$, such as 3:4:5 and 5:12:13
quadrilateral	a four-sided polygon
radius	a line segment inside a circle with one point on the radius and the other point at the center on the circle. The radius is half the diameter. This term can also be used to refer to the length of such a line segment. The plural of *radius* is *radii*.
ray	half of a line. A ray has one endpoint and continues infinitely in one direction. The geometric notation for a ray is with endpoint *A* and passing through point *B* is $\overrightarrow{AB}$.
rectangle	a parallelogram with four right angles
regular polygon	a polygon with all equal sides
rhombus	a parallelogram with four equal sides
right angle	an angle that measures exactly 90°
right triangle	a triangle with an angle that measures exactly 90°
scalene triangle	a triangle with no equal sides
sector	a slice of a circle formed by two radii and an arc
similar polygons	two or more polygons with equal corresponding angles and corresponding sides in proportion
slope	the steepness of a line, as determined by $\frac{\text{vertical change}}{\text{horizontal change}}$, or $\frac{y_2 - y_1}{x_2 - x_1}$, on a coordinate plane where (x_1, y_1) and (x_2, y_2) are two points on that line
solid	a three-dimensional figure
square	a parallelogram with four equal sides and four right angles
supplementary angles	two angles whose sum is 180°

surface area	the sum of the areas of the faces of a solid
tangent	a line that touches a curve (such as a circle) at a single point without cutting across the curve. A tangent line that touches a circle at point *P* is perpendicular to the circle's radius drawn to point *P*
transversal	a line that intersects two or more lines
vertex	a point at which two lines, rays, or line segments connect
vertical angles	two opposite congruent angles formed by intersecting lines
volume	the number of cubic units inside a three-dimensional figure

▶ Formulas

The formulas below for area and volume will be provided to you on the SAT. You do not need to memorize them (although it wouldn't hurt!). Regardless, be sure you understand them thoroughly.

Circle

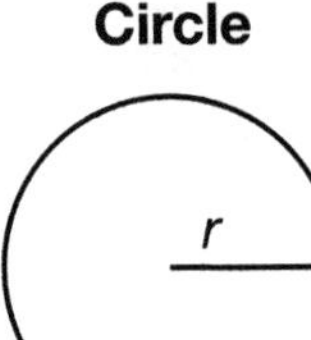

$C = 2\pi r$
$A = \pi r^2$

Rectangle

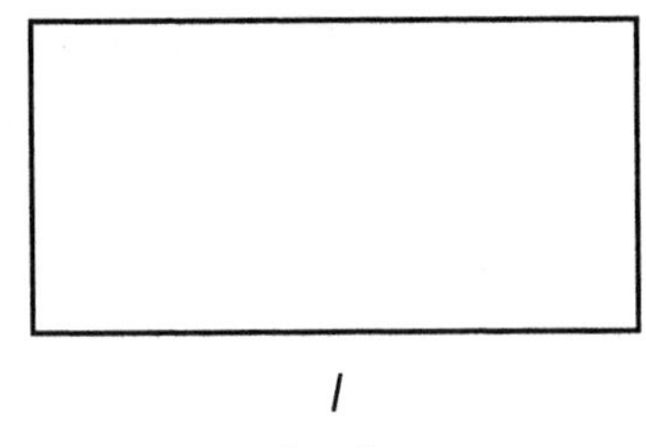

$A = lw$

Triangle

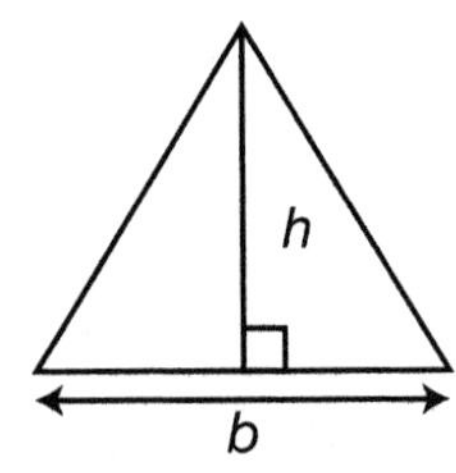

$A = \frac{1}{2}bh$

Cylinder

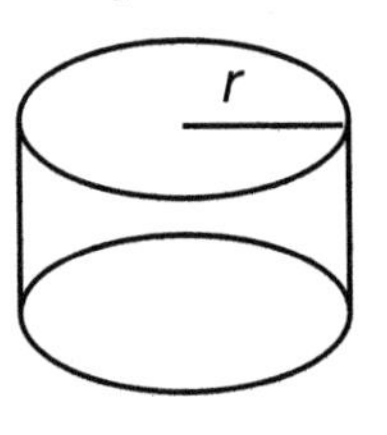

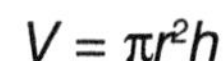

$V = \pi r^2 h$

Rectangle Solid

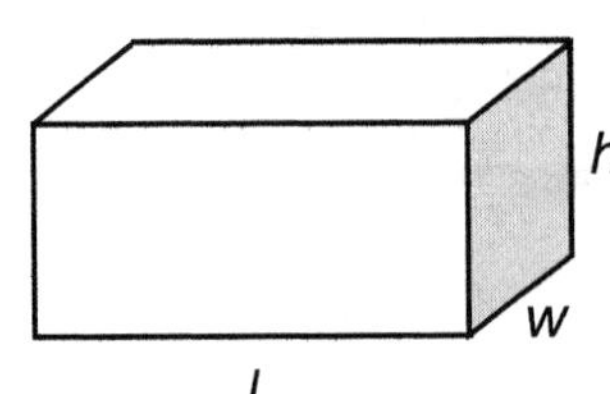

$V = lwh$

C =	Circumference	w =	Width
A =	Area	h =	Height
r =	Radius	V =	Volume
l =	Length	b =	Base

▶ Angles

An **angle** is formed by two rays and an endpoint or line segments that meet at a point, called the **vertex**.

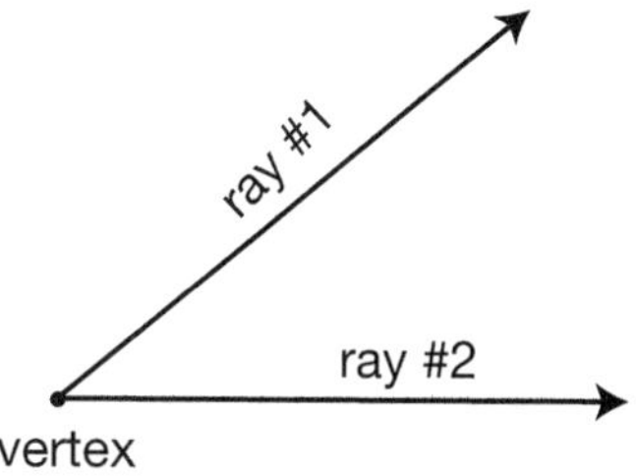

Naming Angles

There are three ways to name an angle.

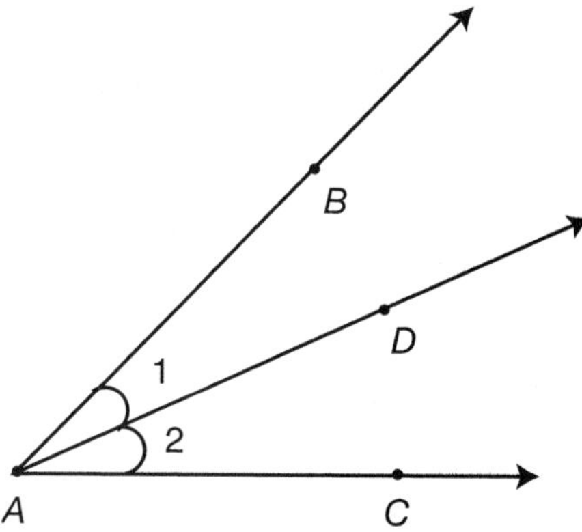

1. An angle can be named by the vertex when no other angles share the same vertex: $\angle A$.
2. An angle can be represented by a number or variable written across from the vertex: $\angle 1$ and $\angle 2$.
3. When more than one angle has the same vertex, three letters are used, with the vertex always being the middle letter: $\angle 1$ can be written as $\angle BAD$ or $\angle DAB$, and $\angle 2$ can be written as $\angle DAC$ or $\angle CAD$.

The Measure of an Angle

The notation $m\angle A$ is used when referring to the measure of an angle (in this case, angle A). For example, if $\angle D$ measures 100°, then $m\angle D = 100°$.

Classifying Angles

Angles are classified into four categories: acute, right, obtuse, and straight.

- An **acute angle** measures less than 90°.

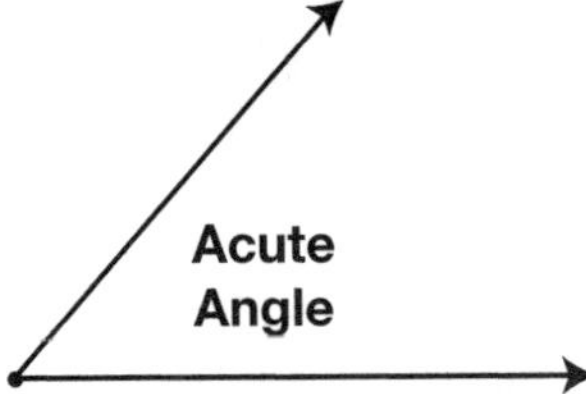

- A **right angle** measures exactly 90°. A right angle is symbolized by a square at the vertex.

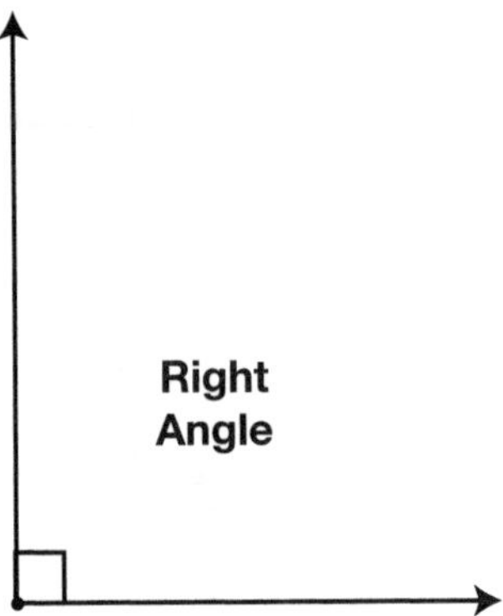

- An **obtuse angle** measures more than 90° but less then 180°.

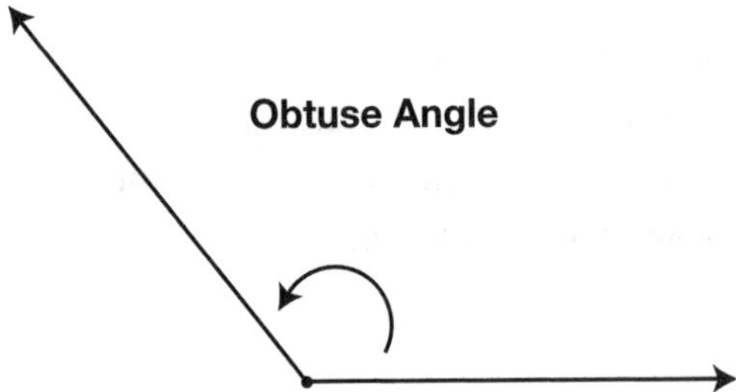

- A **straight angle** measures exactly 180°. A straight angle forms a line.

Practice Question

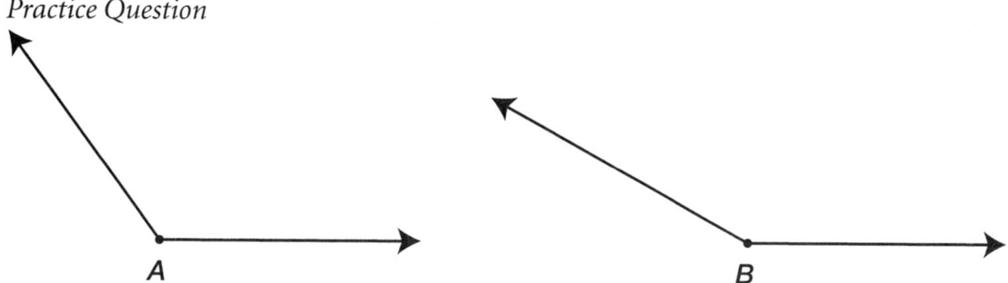

Which of the following must be true about the sum of m∠*A* and m∠*B*?

a. It is equal to 180°.
b. It is less than 180°.
c. It is greater than 180°.
d. It is equal to 360°.
e. It is greater than 360°.

Answer

c. Both ∠*A* and ∠*B* are obtuse, so they are both greater than 90°. Therefore, if 90° + 90° = 180°, then the sum of m∠*A* and m∠*B* must be greater than 180°.

Complementary Angles

Two angles are **complementary** if the sum of their measures is 90°.

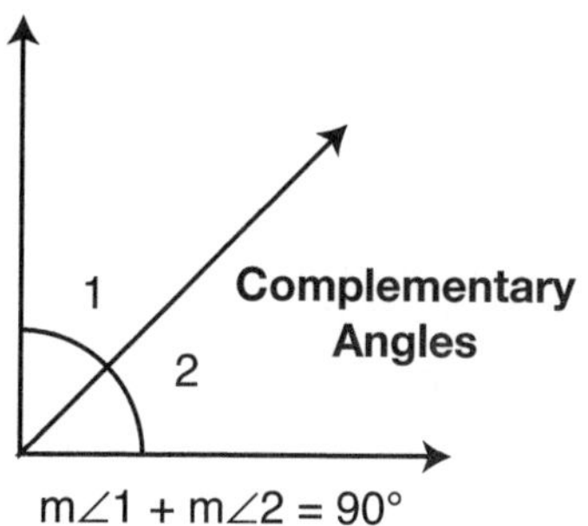

Supplementary Angles

Two angles are **supplementary** if the sum of their measures is 180°.

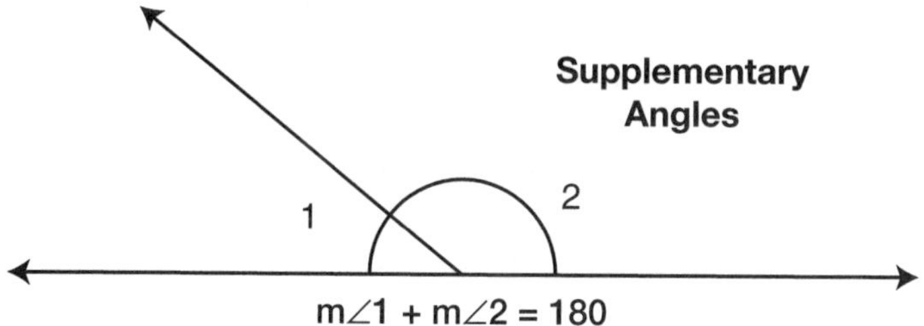

Adjacent angles have the same vertex, share one side, and do not overlap.

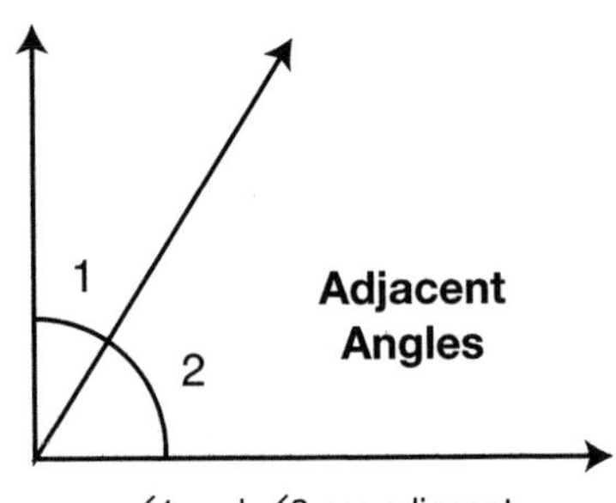

The sum of all adjacent angles around the same vertex is equal to 360°.

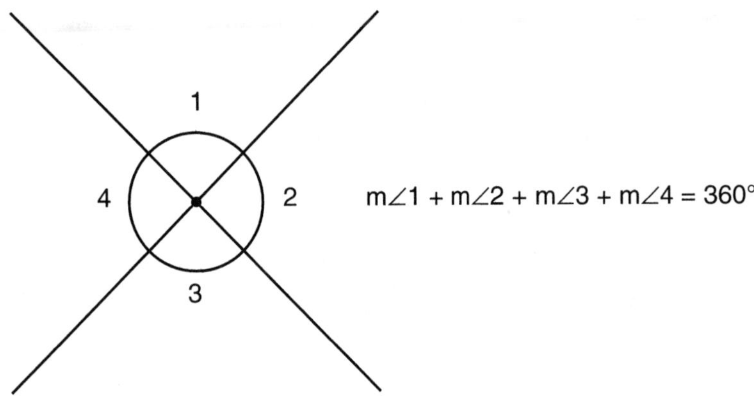

Practice Question

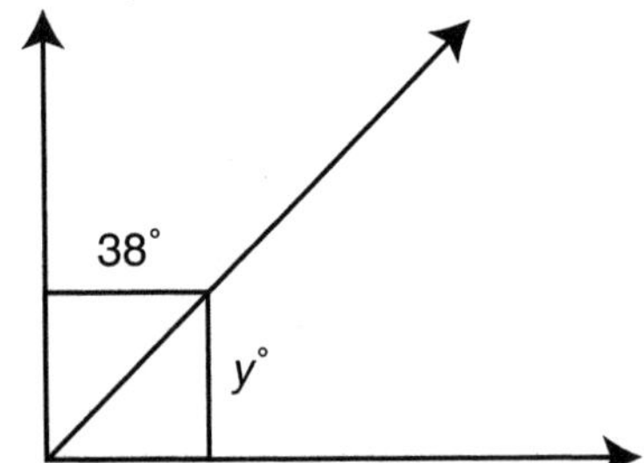

Which of the following must be the value of y?

a. 38
b. 52
c. 90
d. 142
e. 180

Answer

b. The figure shows two complementary angles, which means the sum of the angles equals 90°. If one of the angles is 38°, then the other angle is (90° − 38°). Therefore, $y° = 90° - 38° = 52°$, so $y = 52$.

Angles of Intersecting Lines

When two lines intersect, **vertical angles** are formed. In the figure below, ∠1 and ∠3 are vertical angles and ∠2 and ∠4 are vertical angles.

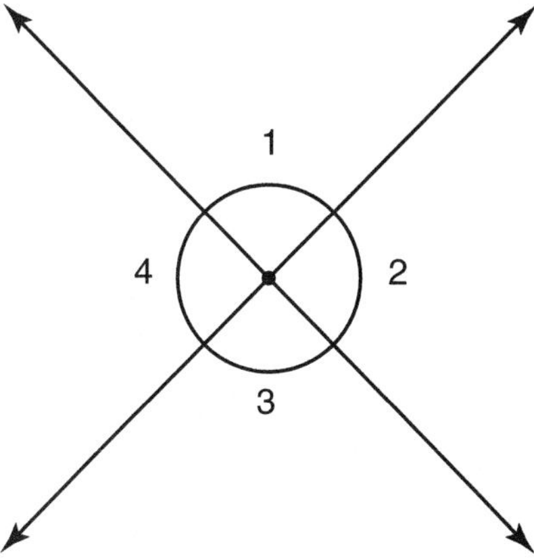

Vertical angles have equal measures:

- m∠1 = m∠3
- m∠2 = m∠4

Vertical angles are supplementary to adjacent angles. The sum of a vertical angle and its adjacent angle is 180°:

- m∠1 + m∠2 = 180°
- m∠2 + m∠3 = 180°
- m∠3 + m∠4 = 180°
- m∠1 + m∠4 = 180°

Practice Question

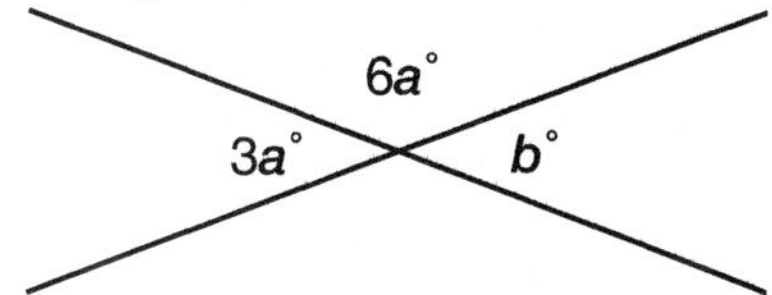

What is the value of *b* in the figure above?

a. 20
b. 30
c. 45
d. 60
e. 120

Answer

d. The drawing shows angles formed by intersecting lines. The laws of intersecting lines tell us that $3a° = b°$ because they are the measures of opposite angles. We also know that $3a° + 6a° = 180°$ because $3a°$ and $6a°$ are measures of supplementary angles. Therefore, we can solve for a:

$3a + 6a = 180$

$9a = 180$

$a = 20$

Because $3a° = b°$, we can solve for b by substituting 20 for a:

$3a = b$

$3(20) = b$

$60 = b$

Bisecting Angles and Line Segments

A line or segment **bisects** a line segment when it divides the second segment into two equal parts.

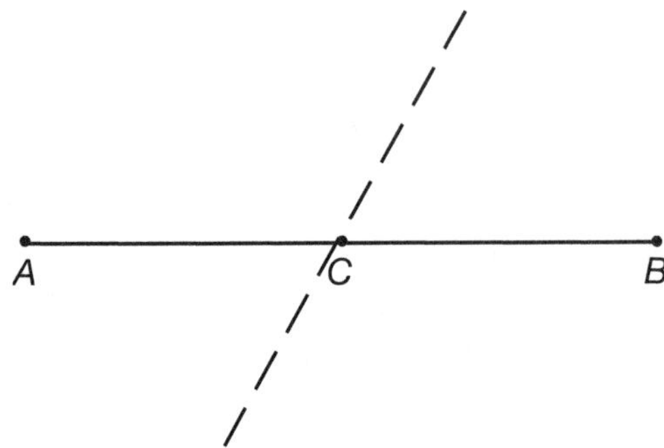

The dotted line **bisects** segment $\overline{AB}$ at point C, so $\overline{AC} = \overline{CB}$.

A line **bisects** an angle when it divides the angle into two equal smaller angles.

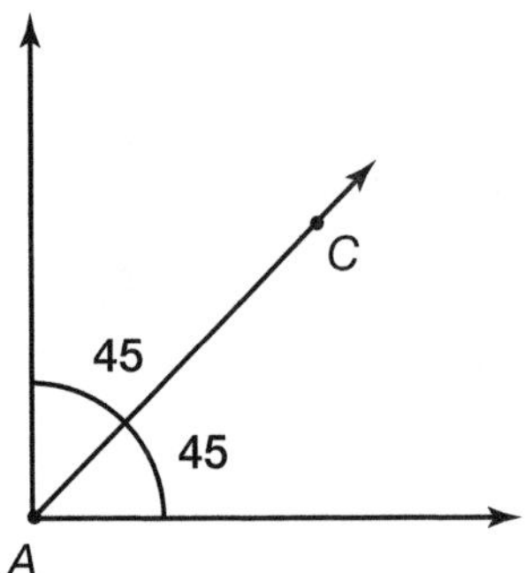

According to the figure, ray $\overrightarrow{AC}$ bisects $\angle A$ because it divides the right angle into two 45° angles.

Angles Formed with Parallel Lines

Vertical angles are the opposite angles formed by the intersection of any two lines. In the figure below, ∠1 and ∠3 are vertical angles because they are opposite each other. ∠2 and ∠4 are also vertical angles.

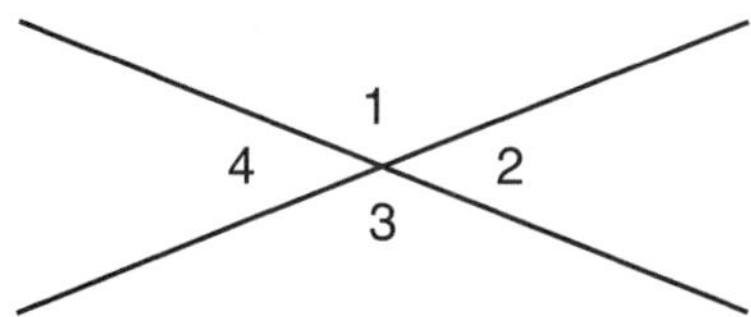

A special case of vertical angles occurs when a transversal line intersects two parallel lines.

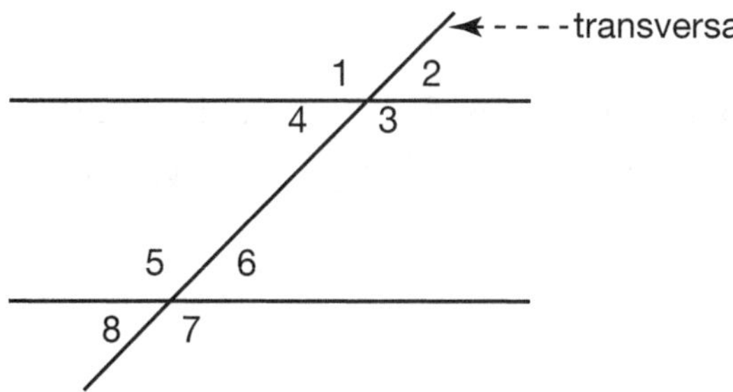

The following rules are true when a transversal line intersects two parallel lines.

- There are four sets of vertical angles:
  ∠1 and ∠3
  ∠2 and ∠4
  ∠5 and ∠7
  ∠6 and ∠8
- Four of these vertical angles are obtuse:
  ∠1, ∠3, ∠5, and ∠7
- Four of these vertical angles are acute:
  ∠2, ∠4, ∠6, and ∠8
- The obtuse angles are equal:
  ∠1 = ∠3 = ∠5 = ∠7
- The acute angles are equal:
  ∠2 = ∠4 = ∠6 = ∠8
- In this situation, any acute angle added to any obtuse angle is supplementary.
  m∠1 + m∠2 = 180°
  m∠2 + m∠3 = 180°
  m∠3 + m∠4 = 180°
  m∠1 + m∠4 = 180°
  m∠5 + m∠6 = 180°
  m∠6 + m∠7 = 180°
  m∠7 + m∠8 = 180°
  m∠5 + m∠8 = 180°

You can use these rules of vertical angles to solve problems.

Example

In the figure below, if $c \parallel d$, what is the value of x?

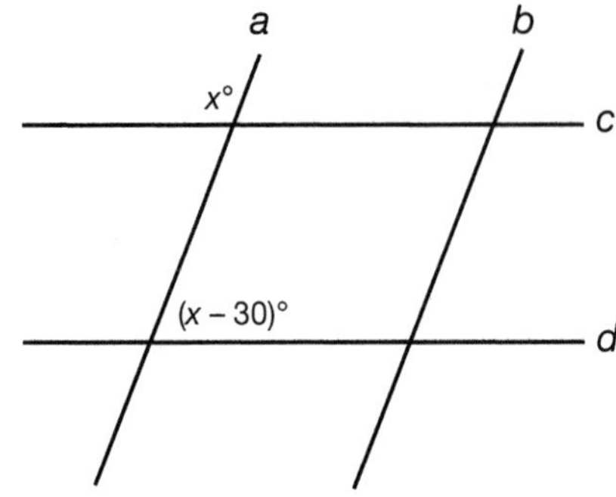

Because $c \parallel d$, you know that the sum of an acute angle and an obtuse angle formed by an intersecting line (line a) is equal to 180°. $\angle x$ is obtuse and $\angle(x - 30)$ is acute, so you can set up the equation $x + (x - 30) = 180$. Now solve for x:

$x + (x - 30) = 180$

$2x - 30 = 180$

$2x - 30 + 30 = 180 + 30$

$2x = 210$

$x = 105$

Therefore, $m\angle x = 105°$. The acute angle is equal to $180 - 105 = 75°$.

Practice Question

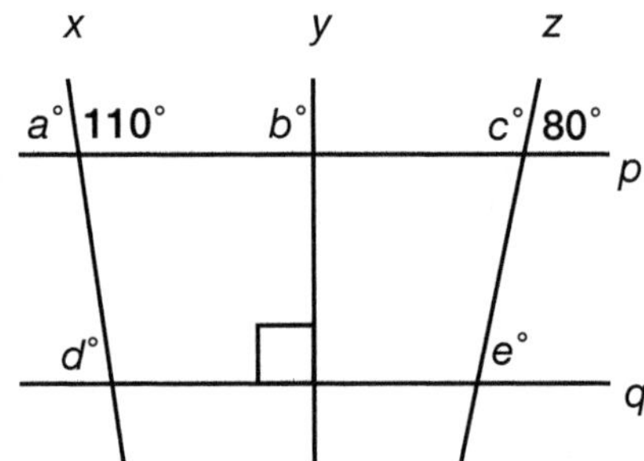

If $p \parallel q$, which the following is equal to 80?

a. a
b. b
c. c
d. d
e. e

Answer

e. Because $p \parallel q$, the angle with measure 80° and the angle with measure $e°$ are corresponding angles, so they are equivalent. Therefore $e° = 80°$, and $e = 80$.

Interior and Exterior Angles

Exterior angles are the angles on the outer sides of two lines intersected by a transversal. **Interior angles** are the angles on the inner sides of two lines intersected by a transversal.

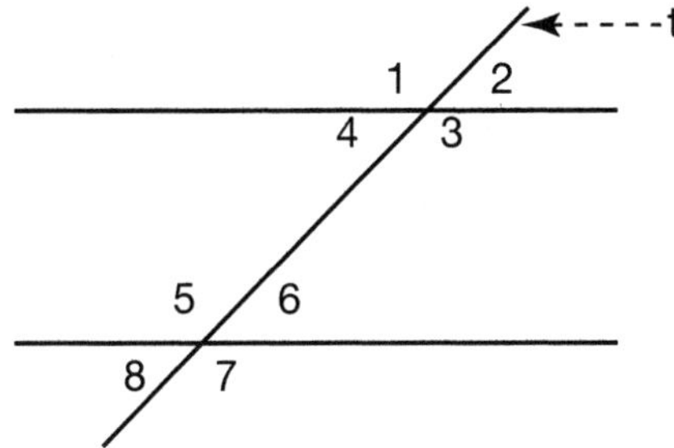

In the figure above:

$\angle 1$, $\angle 2$, $\angle 7$, and $\angle 8$ are exterior angles.
$\angle 3$, $\angle 4$, $\angle 5$, and $\angle 6$ are interior angles.

▶ Triangles

Angles of a Triangle

The measures of the three angles in a triangle always add up to 180°.

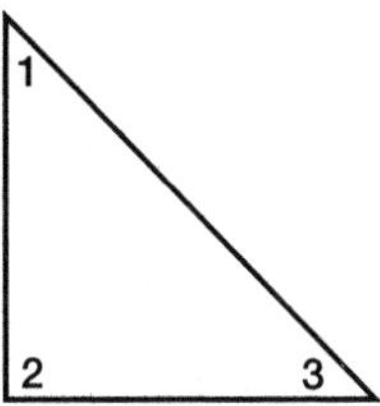

$m\angle 1 + m\angle 2 + m\angle 3 = 180°$

Exterior Angles of a Triangle

Triangles have three exterior angles. $\angle a$, $\angle b$, and $\angle c$ are the exterior angles of the triangle below.

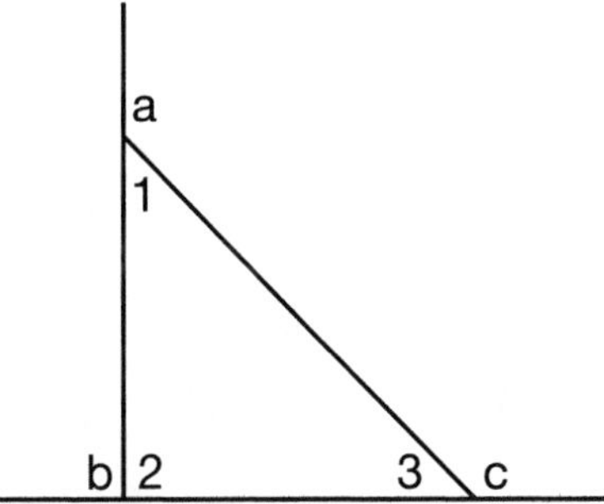

- An exterior angle and interior angle that share the same vertex are supplementary:

$m\angle 1 + m\angle a = 180°$
$m\angle 2 + m\angle b = 180°$
$m\angle 3 + m\angle c = 180°$

- An exterior angle is equal to the sum of the non-adjacent interior angles:
  $m\angle a = m\angle 2 + m\angle 3$
  $m\angle b = m\angle 1 + m\angle 3$
  $m\angle c = m\angle 1 + m\angle 2$

The sum of the exterior angles of any triangle is 360°.

Practice Question

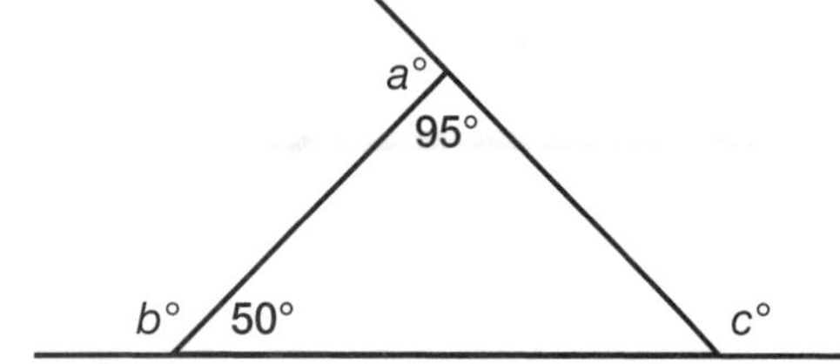

Based on the figure, which of the following must be true?

I. $a < b$
II. $c = 135°$
III. $b < c$

a. I only
b. III only
c. I and III only
d. II and III only
e. I, II, and III

Answer

c. To solve, you must determine the value of the third angle of the triangle and the values of a, b, and c. The third angle of the triangle = 180° − 95° − 50° = 35° (because the sum of the measures of the angles of a triangle are 180°).
$a = 180 - 95 = 85$ (because $\angle a$ and the angle that measures 95° are supplementary).
$b = 180 - 50 = 130$ (because $\angle b$ and the angle that measures 50° are supplementary).
$c = 180 - 35 = 145$ (because $\angle c$ and the angle that measures 35° are supplementary).
Now we can evaluate the three statements:
I: $a < b$ is TRUE because $a = 85$ and $b = 130$.
II: $c = 135°$ is FALSE because $c = 145°$.
III: $b < c$ is TRUE because $b = 130$ and $c = 145$.
Therefore, only I and III are true.

Types of Triangles

You can classify triangles into three categories based on the number of equal sides.

- **Scalene Triangle:** no equal sides

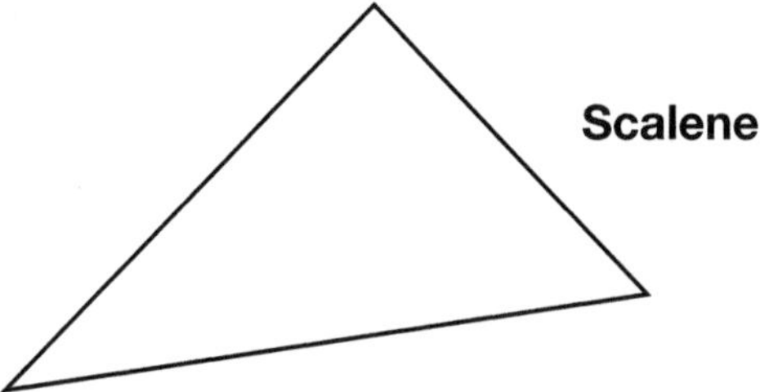

- **Isosceles Triangle:** two equal sides

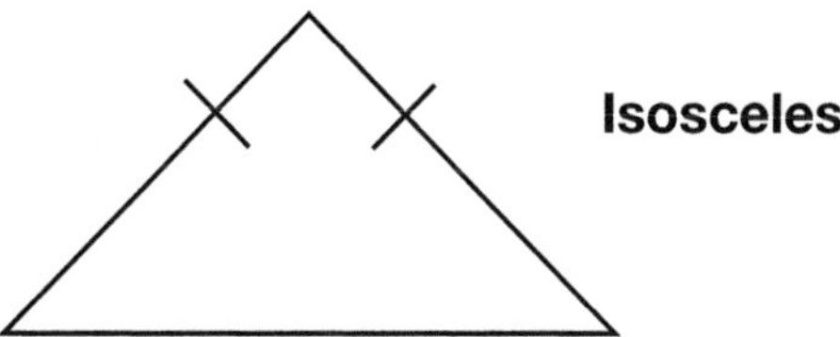

- **Equilateral Triangle:** all equal sides

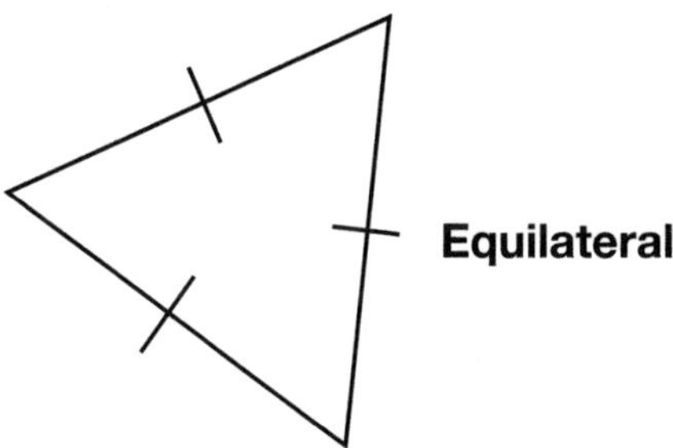

You also can classify triangles into three categories based on the measure of the greatest angle:

- **Acute Triangle:** greatest angle is acute

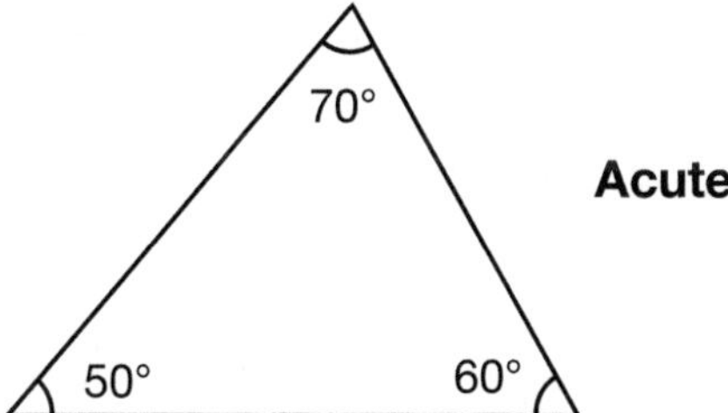

- **Right Triangle:** greatest angle is 90°

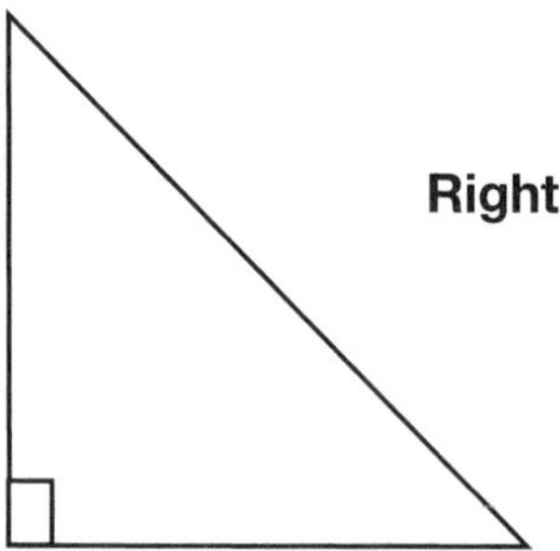

- **Obtuse Triangle:** greatest angle is obtuse

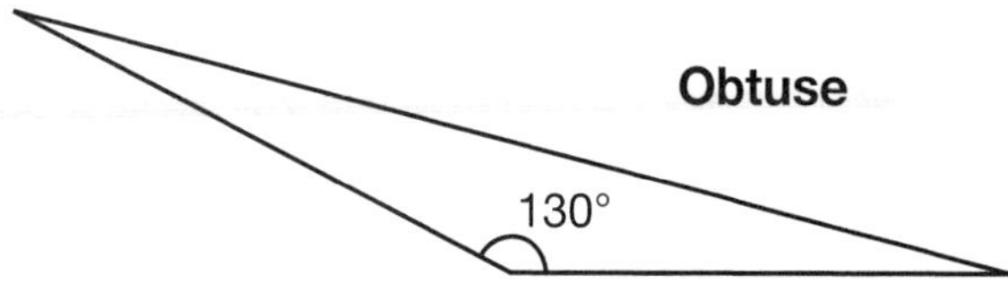

Angle-Side Relationships

Understanding the angle-side relationships in isosceles, equilateral, and right triangles is essential in solving questions on the SAT.

- In **isosceles triangles**, equal angles are opposite equal sides.

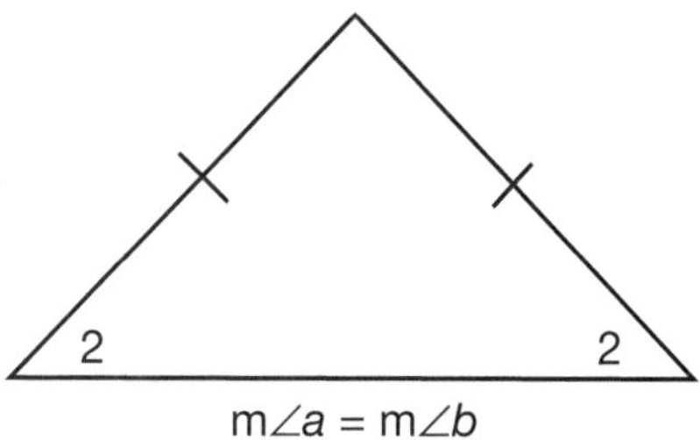

- In **equilateral triangles**, all sides are equal and all angles are 60°.

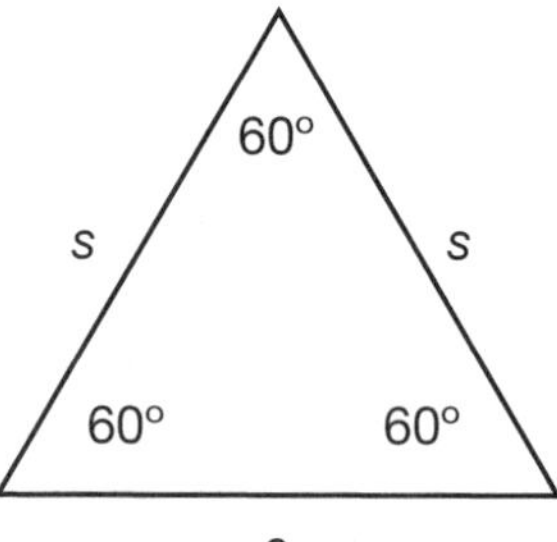

- In **right triangles**, the side opposite the right angle is called the hypotenuse.

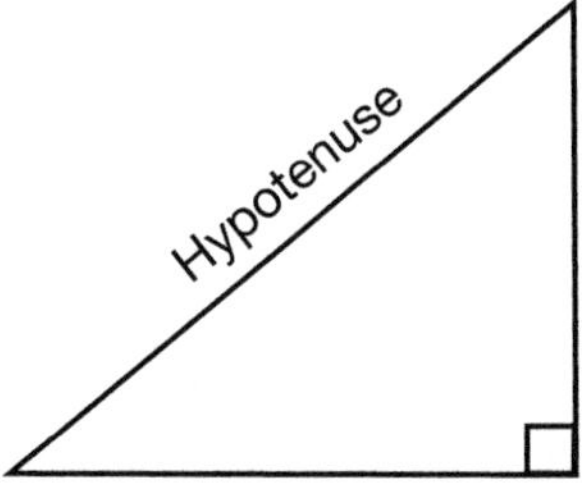

Practice Question

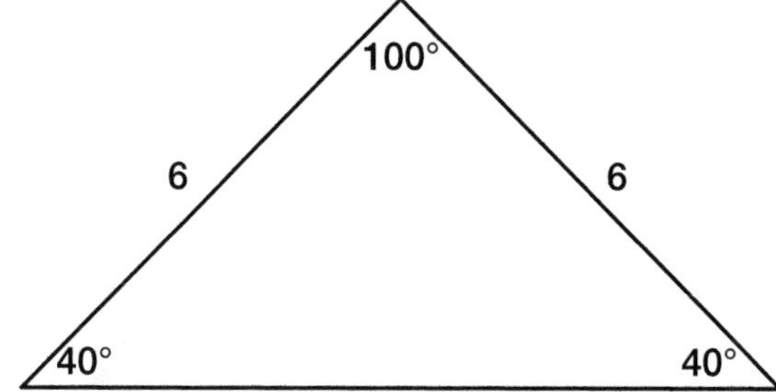

Which of the following best describes the triangle above?

a. scalene and obtuse
b. scalene and acute
c. isosceles and right
d. isosceles and obtuse
e. isosceles and acute

Answer

d. The triangle has an angle greater than 90°, which makes it *obtuse.* Also, the triangle has two equal sides, which makes it *isosceles.*

▶ Pythagorean Theorem

The **Pythagorean theorem** is an important tool for working with right triangles. It states:

$a^2 + b^2 = c^2$, where a and b represent the lengths of the *legs* and c represents the length of the *hypotenuse* of a right triangle.

Therefore, if you know the lengths of two sides of a right triangle, you can use the Pythagorean theorem to determine the length of the third side.

Example

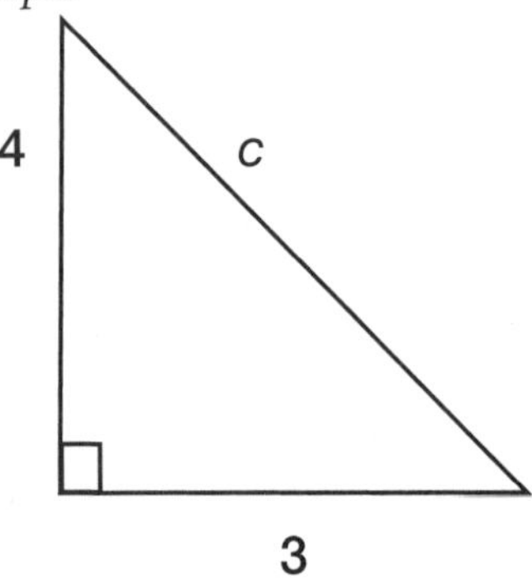

$a^2 + b^2 = c^2$

$3^2 + 4^2 = c^2$

$9 + 16 = c^2$

$25 = c^2$

$\sqrt{25} = \sqrt{c^2}$

$5 = c$

Example

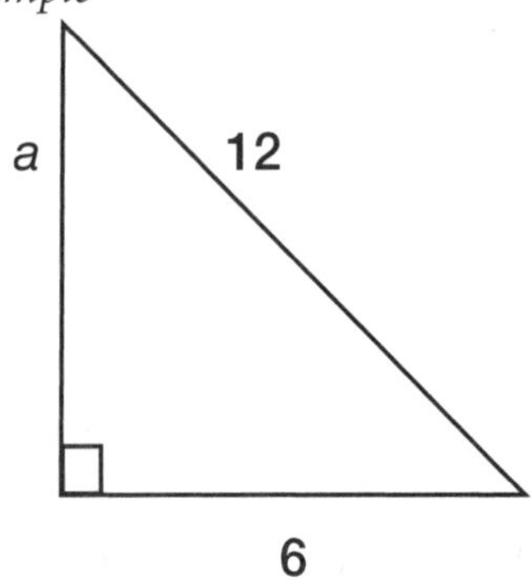

$a^2 + b^2 = c^2$

$a^2 + 6^2 = 12^2$

$a^2 + 36 = 144$

$a^2 + 36 - 36 = 144 - 36$

$a^2 = 108$

$\sqrt{a^2} = \sqrt{108}$

$a = \sqrt{108}$

Practice Question

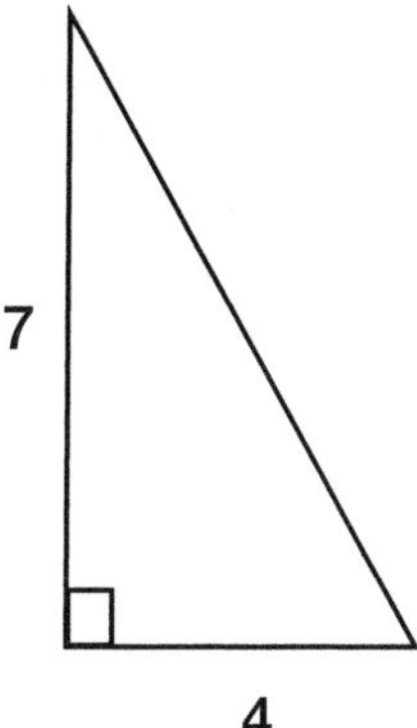

What is the length of the hypotenuse in the triangle above?

a. $\sqrt{11}$
b. 8
c. $\sqrt{65}$
d. 11
e. 65

Answer

c. Use the Pythagorean theorem: $a^2 + b^2 = c^2$, where $a = 7$ and $b = 4$.

$a^2 + b^2 = c^2$
$7^2 + 4^2 = c^2$
$49 + 16 = c^2$
$65 = c^2$
$\sqrt{65} = \sqrt{c^2}$
$\sqrt{65} = c$

Pythagorean Triples

A **Pythagorean triple** is a set of three positive integers that satisfies the Pythagorean theorem, $a^2 + b^2 = c^2$.

Example

The set 3:4:5 is a Pythagorean triple because:
$3^2 + 4^2 = 5^2$
$9 + 16 = 25$
$25 = 25$

Multiples of Pythagorean triples are also Pythagorean triples.

Example

Because set 3:4:5 is a Pythagorean triple, 6:8:10 is also a Pythagorean triple:
$6^2 + 8^2 = 10^2$
$36 + 64 = 100$
$100 = 100$

Pythagorean triples are important because they help you identify right triangles and identify the lengths of the sides of right triangles.

Example

What is the measure of $\angle a$ in the triangle below?

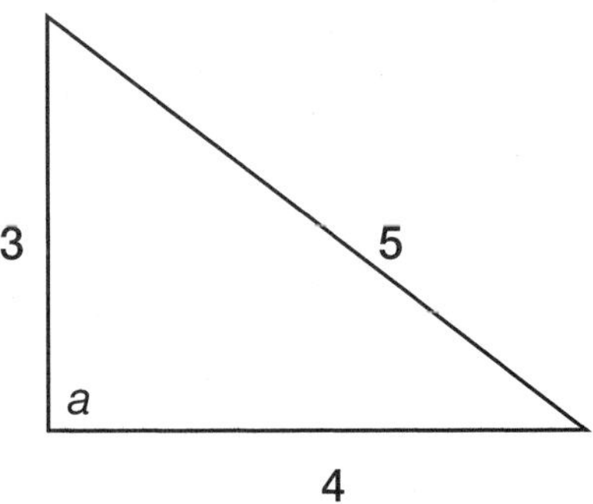

Because this triangle shows a Pythagorean triple (3:4:5), you know it is a right triangle. Therefore, $\angle a$ must measure 90°.

Example

A right triangle has a leg of 8 and a hypotenuse of 10. What is the length of the other leg?

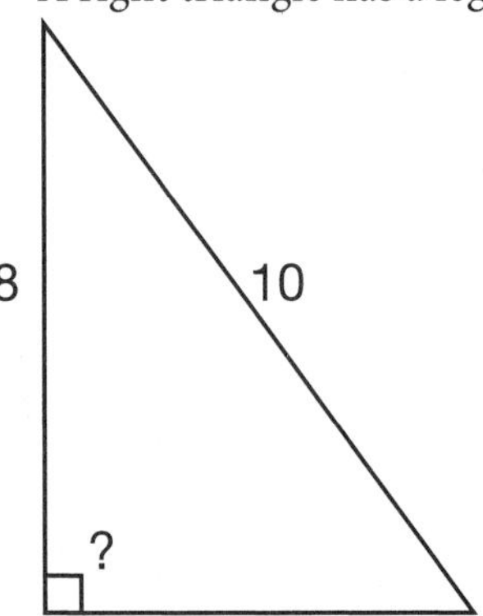

Because this triangle is a right triangle, you know its measurements obey the Pythagorean theorem. You could plug 8 and 10 into the formula and solve for the missing leg, but you don't have to. The triangle shows two parts of a Pythagorean triple (?:8:10), so you know that the missing leg must complete the triple. Therefore, the second leg has a length of 6.

It is useful to memorize a few of the smallest Pythagorean triples:

3:4:5	$3^2 + 4^2 = 5^2$
6:8:10	$6^2 + 8^2 = 10^2$
5:12:13	$5^2 + 12^2 = 13^2$
7:24:25	$7^2 + 24^2 = 25^2$
8:15:17	$8^2 + 15^2 = 17^2$

Practice Question

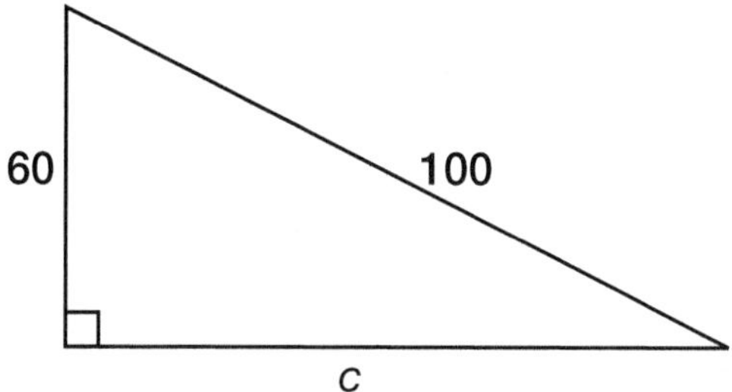

What is the length of c in the triangle above?

a. 30
b. 40
c. 60
d. 80
e. 100

Answer

d. You could use the Pythagorean theorem to solve this question, but if you notice that the triangle shows two parts of a Pythagorean triple, you don't have to. 60:c:100 is a multiple of 6:8:10 (which is a multiple of 3:4:5). Therefore, c must equal 80 because 60:80:100 is the same ratio as 6:8:10.

45-45-90 Right Triangles

An **isosceles right triangle** is a right triangle with two angles each measuring 45°.

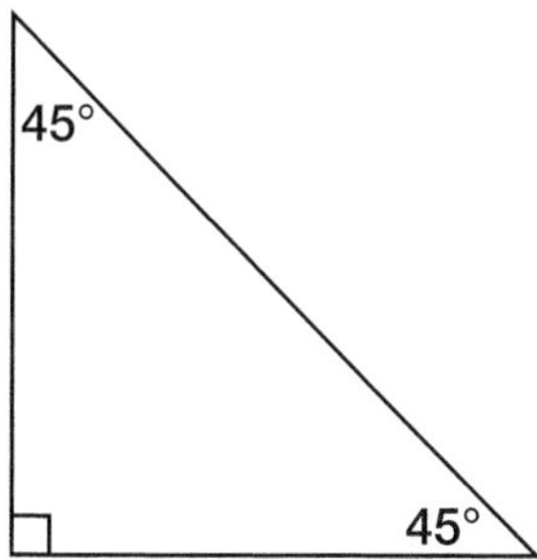

Special rules apply to isosceles right triangles:

- the length of the hypotenuse = $\sqrt{2} \times$ the length of a leg of the triangle

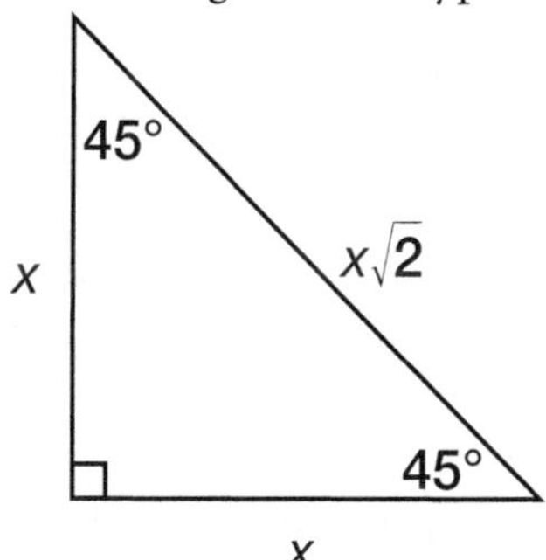

- the length of each leg is $\frac{\sqrt{2}}{2}$ × the length of the hypotenuse

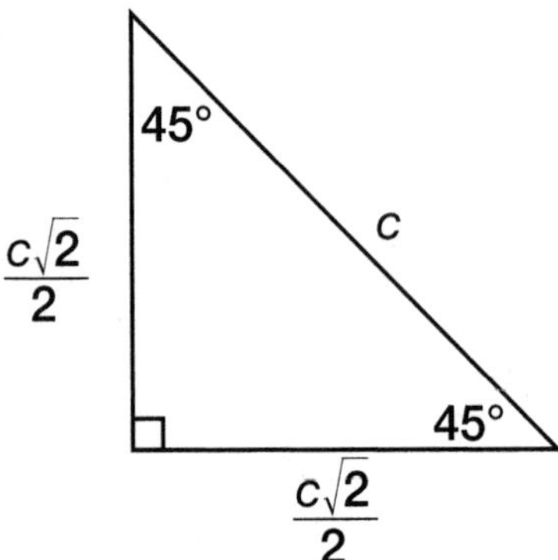

You can use these special rules to solve problems involving isosceles right triangles.

Example

In the isosceles right triangle below, what is the length of a leg, *x*?

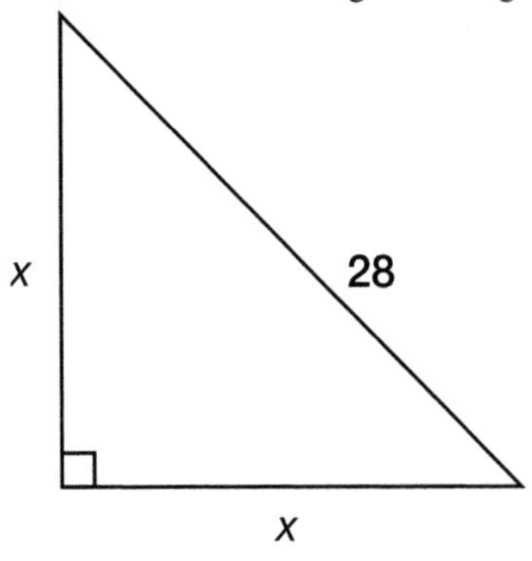

$x = \frac{\sqrt{2}}{2} \times$ the length of the hypotenuse

$x = \frac{\sqrt{2}}{2} \times 28$

$x = \frac{28\sqrt{2}}{2}$

$x = 14\sqrt{2}$

Practice Question

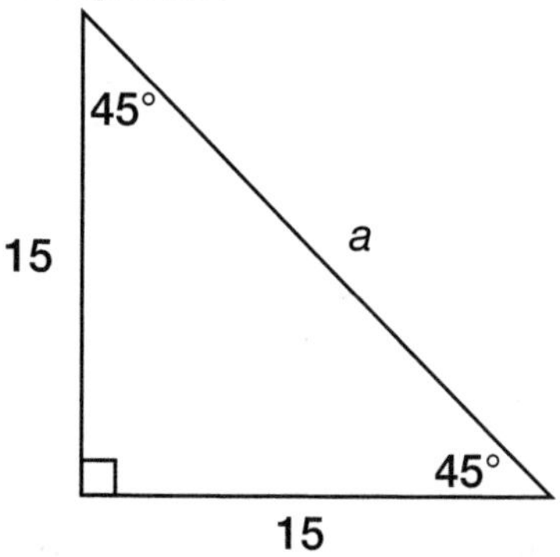

What is the length of *a* in the triangle above?

a. $\frac{15\sqrt{2}}{4}$

b. $\frac{15\sqrt{2}}{2}$

c. $15\sqrt{2}$

d. 30

e. $30\sqrt{2}$

Answer

c. In an isosceles right triangle, the length of the hypotenuse = $\sqrt{2}$ × the length of a leg of the triangle. According to the figure, one leg = 15. Therefore, the hypotenuse is $15\sqrt{2}$.

30-60-90 Triangles

Special rules apply to right triangles with one angle measuring 30° and another angle measuring 60°.

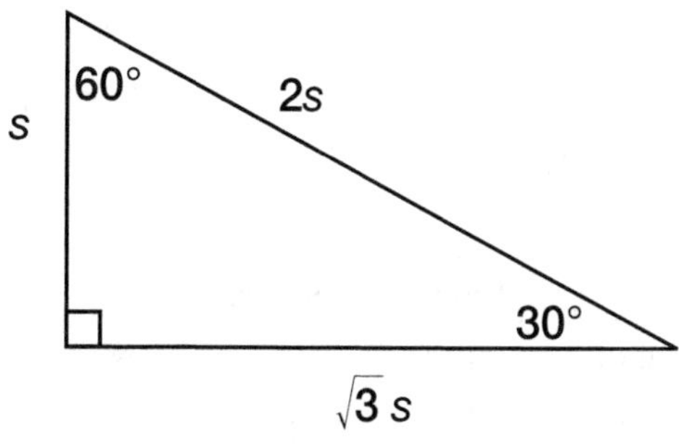

- the hypotenuse = 2 × the length of the leg opposite the 30° angle
- the leg opposite the 30° angle = $\frac{1}{2}$ × the length of the hypotenuse
- the leg opposite the 60° angle = $\sqrt{3}$ × the length of the other leg

You can use these rules to solve problems involving 30-60-90 triangles.

Example

What are the lengths of x and y in the triangle below?

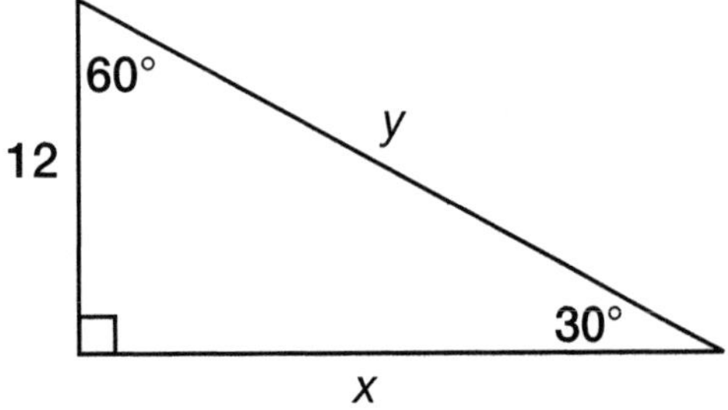

The hypotenuse = 2 × the length of the leg opposite the 30° angle. Therefore, you can write an equation:

$y = 2 \times 12$

$y = 24$

The leg opposite the 60° angle = $\sqrt{3}$ × the length of the other leg. Therefore, you can write an equation:

$x = 12\sqrt{3}$

Practice Question

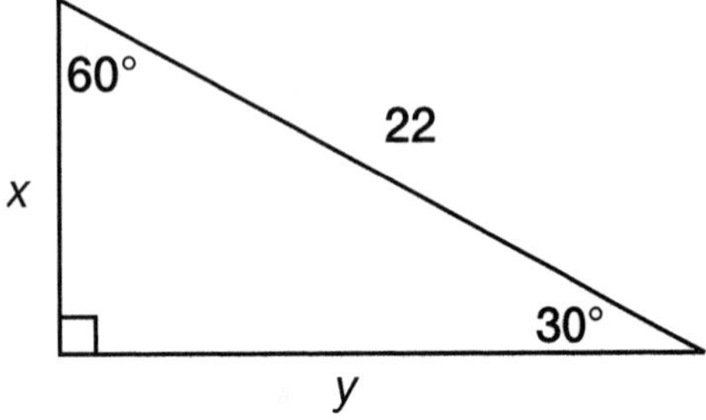

What is the length of y in the triangle above?

a. 11
b. $11\sqrt{2}$
c. $11\sqrt{3}$
d. $22\sqrt{2}$
e. $22\sqrt{3}$

Answer

c. In a 30-60-90 triangle, the leg opposite the 30° angle = half the length of the hypotenuse. The hypotenuse is 22, so the leg opposite the 30° angle = 11. The leg opposite the 60° angle = $\sqrt{3}$ × the length of the other leg. The other leg = 11, so the leg opposite the 60° angle = $11\sqrt{3}$.

Triangle Trigonometry

There are special ratios we can use when working with right triangles. They are based on the trigonometric functions called **sine**, **cosine**, and **tangent**.

For an angle, Θ, within a right triangle, we can use these formulas:

$\sin \Theta = \frac{\text{opposite}}{\text{hypotenuse}}$ $\cos \Theta = \frac{\text{adjacent}}{\text{hypotenuse}}$ $\tan \Theta = \frac{\text{opposite}}{\text{adjacent}}$

To find sin Θ...

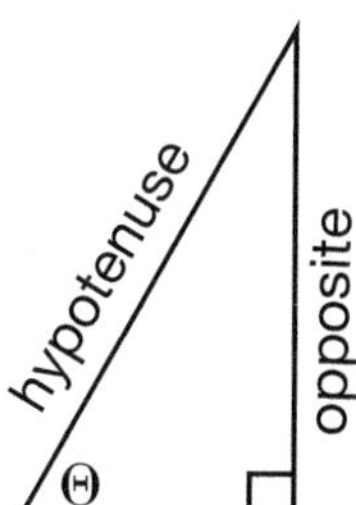

To find cos Θ...

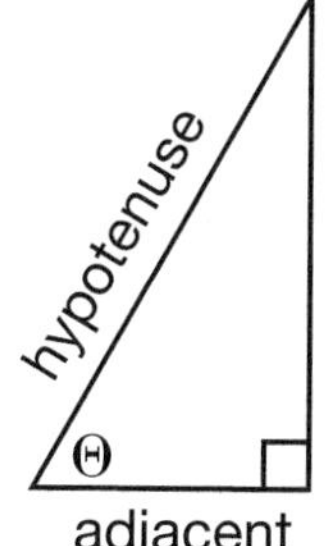

To find tan Θ...

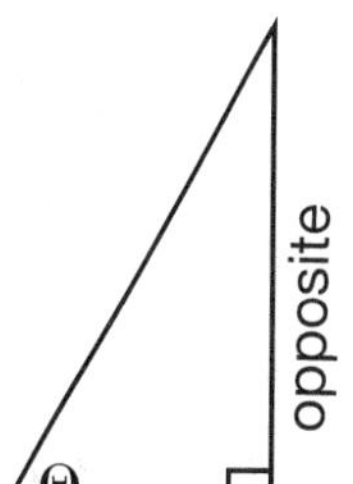

The popular mnemonic to use to remember these formulas is **SOH CAH TOA.**

SOH stands for **S**in: **O**pposite/**H**ypotenuse
CAH stands for **C**os: **A**djacent/**H**ypotenuse
TOA stands for **T**an: **O**pposite/**A**djacent

TRIG VALUES OF SOME COMMON ANGLES

	SIN	COS	TAN
30°	$\frac{1}{2}$	$\frac{\sqrt{3}}{2}$	$\frac{\sqrt{3}}{3}$
45°	$\frac{\sqrt{2}}{2}$	$\frac{\sqrt{2}}{2}$	1
60°	$\frac{\sqrt{3}}{2}$	$\frac{1}{2}$	$\sqrt{3}$

Although trigonometry is tested on the SAT, all SAT trigonometry questions can also be solved using geometry (such as rules of 45-45-90 and 30-60-90 triangles), so knowledge of trigonometry is not essential. But if you don't bother learning trigonometry, be sure you understand triangle geometry completely.

Example

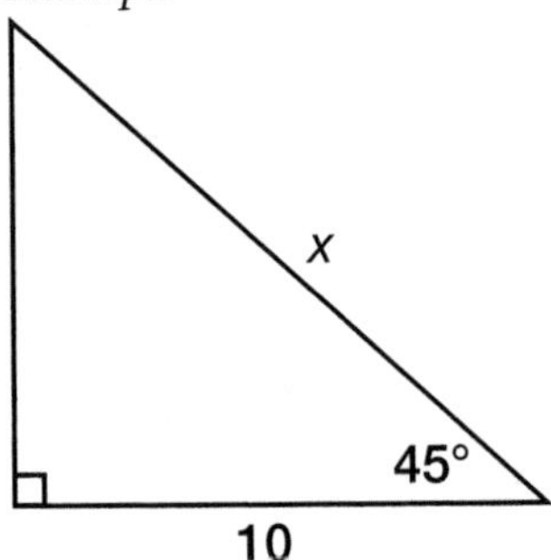

First, let's solve using trigonometry:
We know that $\cos 45° = \frac{\sqrt{2}}{2}$, so we can write an equation:

$\frac{\text{adjacent}}{\text{hypotenuse}} = \frac{\sqrt{2}}{2}$

$\frac{10}{x} = \frac{\sqrt{2}}{2}$ Find cross products.

$2 \times 10 = x\sqrt{2}$ Simplify.

$20 = x\sqrt{2}$

$\frac{20}{\sqrt{2}} = x$

Now, multiply $\frac{20}{\sqrt{2}}$ by $\frac{\sqrt{2}}{\sqrt{2}}$ (which equals 1), to remove the $\sqrt{2}$ from the denominator.

$\frac{\sqrt{2}}{\sqrt{2}} \times \frac{20}{\sqrt{2}} = x$

$\frac{20\sqrt{2}}{2} = x$

$10\sqrt{2} = x$

Now let's solve using rules of 45-45-90 triangles, which is a lot simpler:
The length of the hypotenuse = $\sqrt{2} \times$ the length of a leg of the triangle. Therefore, because the leg is 10, the hypotenuse is $\sqrt{2} \times 10 = 10\sqrt{2}$.

▶ Circles

A **circle** is a closed figure in which each point of the circle is the same distance from the center of the circle.

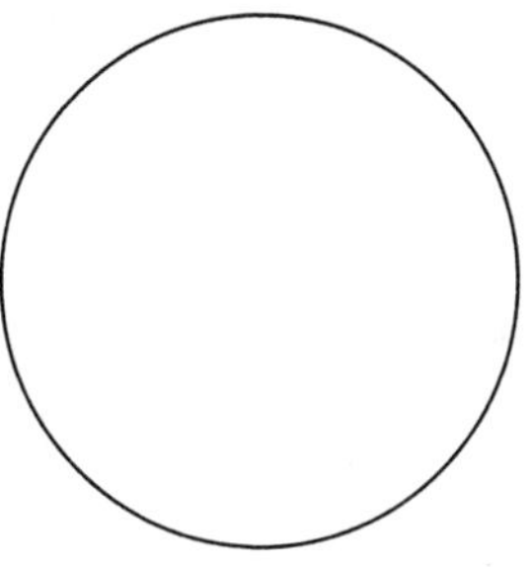

Angles and Arcs of a Circle

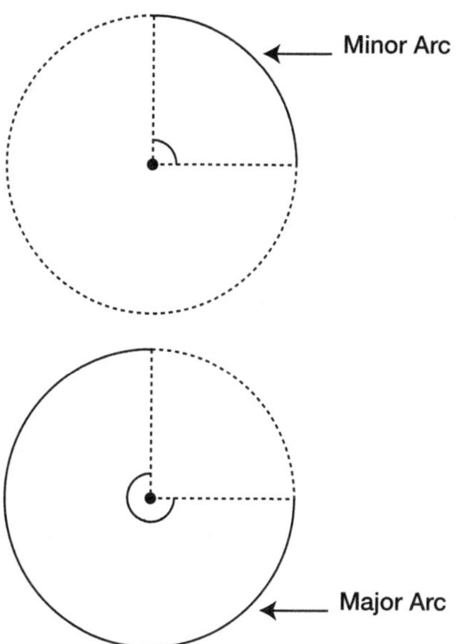

- An **arc** is a curved section of a circle.
- A **minor arc** is an arc less than or equal to 180°. A **major arc** is an arc greater than or equal to 180°.

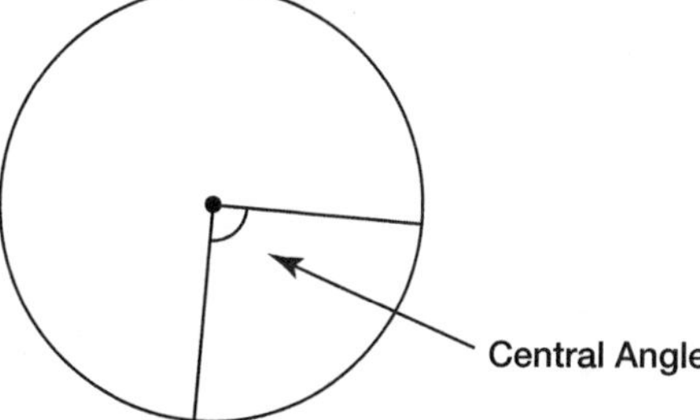

- A **central angle** of a circle is an angle with its vertex at the center and sides that are radii. Arcs have the same degree measure as the central angle whose sides meet the circle at the two ends of the arc.

Length of an Arc

To find the length of an arc, multiply the circumference of the circle, $2\pi r$, where r = the radius of the circle, by the fraction $\frac{x}{360}$, with x being the degree measure of the central angle:

$$2\pi r \times \frac{x}{360} = \frac{2\pi rx}{360} = \frac{\pi rx}{180}$$

Example

Find the length of the arc if $x = 90$ and $r = 56$.

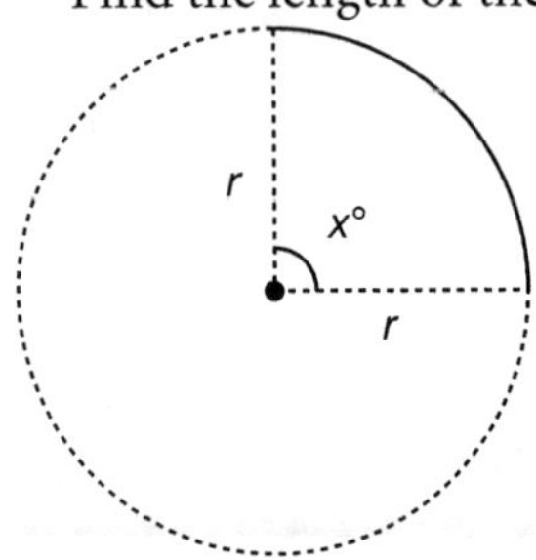

$L = \frac{\pi rx}{180}$

$L = \frac{\pi(56)(90)}{180}$

$L = \frac{\pi(56)}{2}$

$L = 28\pi$

The length of the arc is 28π.

Practice Question

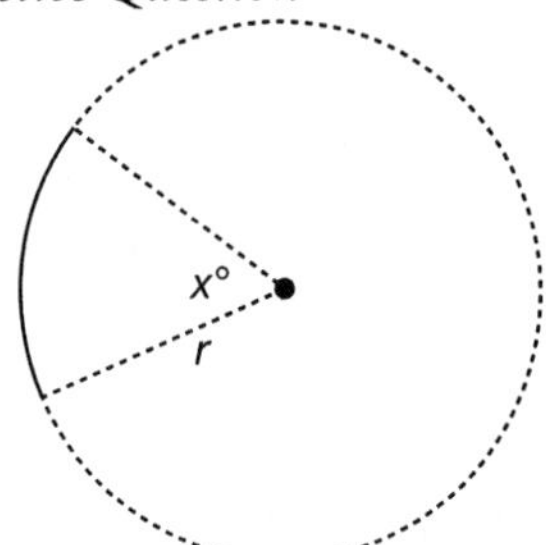

If $x = 32$ and $r = 18$, what is the length of the arc shown in the figure above?

a. $\frac{16\pi}{5}$

b. $\frac{32\pi}{5}$

c. 36π

d. $\frac{288\pi}{5}$

e. 576π

Answer

a. To find the length of an arc, use the formula $\frac{\pi rx}{180}$, where $r =$ the radius of the circle and $x =$ the measure of the central angle of the arc. In this case, $r = 18$ and $x = 32$.

$\frac{\pi rx}{180} = \frac{\pi(18)(32)}{180} = \frac{\pi(32)}{10} = \frac{\pi(16)}{5} = \frac{16\pi}{5}$

Area of a Sector

A **sector** of a circle is a slice of a circle formed by two radii and an arc.

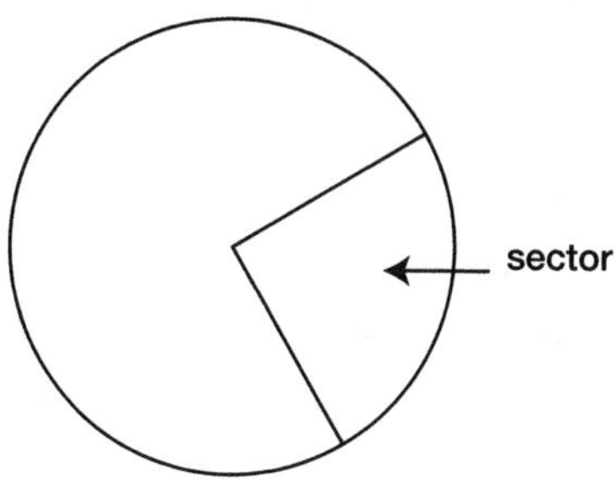

To find the area of a sector, multiply the area of a circle, πr^2, by the fraction $\frac{x}{360}$, with x being the degree measure of the central angle: $\frac{\pi r^2 x}{360}$.

Example

Given $x = 120$ and $r = 9$, find the area of the sector:

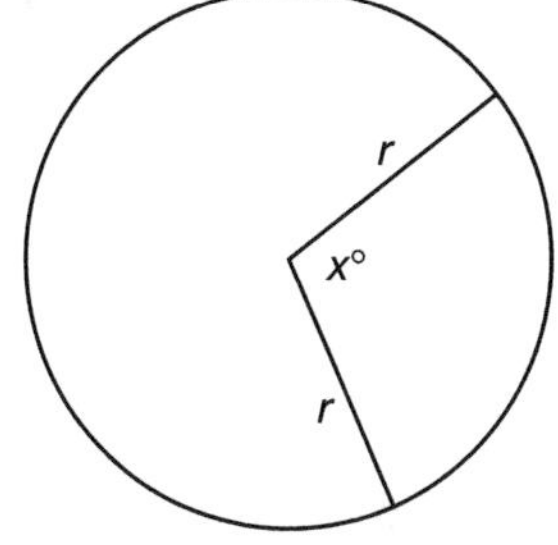

$A = \frac{\pi r^2 x}{360}$

$A = \frac{\pi(9^2)(120)}{360}$

$A = \frac{\pi(9^2)}{3}$

$A = \frac{81\pi}{3}$

$A = 27\pi$

The area of the sector is 27π.

Practice Question

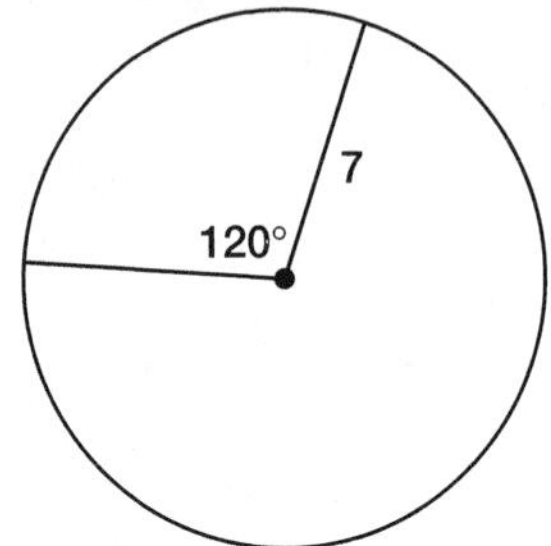

What is the area of the sector shown above?

a. $\frac{49\pi}{360}$
b. $\frac{7\pi}{3}$
c. $\frac{49\pi}{3}$
d. 280π
e. $5{,}880\pi$

Answer

c. To find the area of a sector, use the formula $\frac{\pi r^2 x}{360}$, where r = the radius of the circle and x = the measure of the central angle of the arc. In this case, $r = 7$ and $x = 120$.

$\frac{\pi r^2 x}{360} = \frac{\pi(7^2)(120)}{360} = \frac{\pi(49)(120)}{360} = \frac{\pi(49)}{3} = \frac{49\pi}{3}$

Tangents

A **tangent** is a line that intersects a circle at one point only.

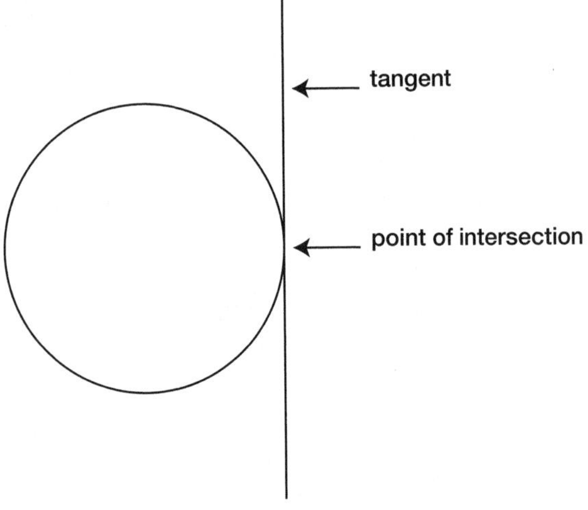

There are two rules related to tangents:

1. A radius whose endpoint is on the tangent is always perpendicular to the tangent line.

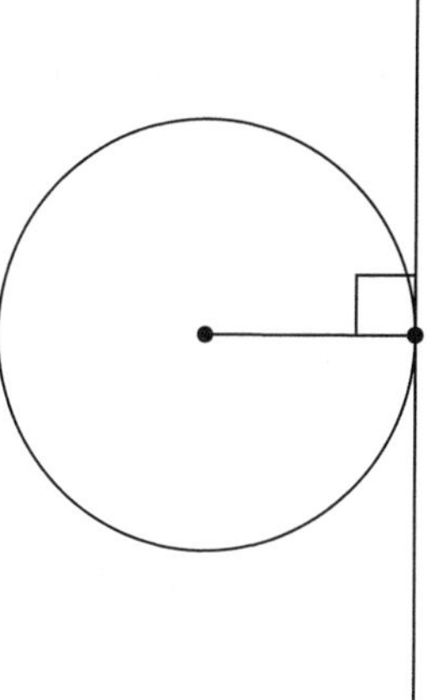

2. Any point outside a circle can extend exactly two tangent lines to the circle. The distances from the origin of the tangents to the points where the tangents intersect with the circle are equal.

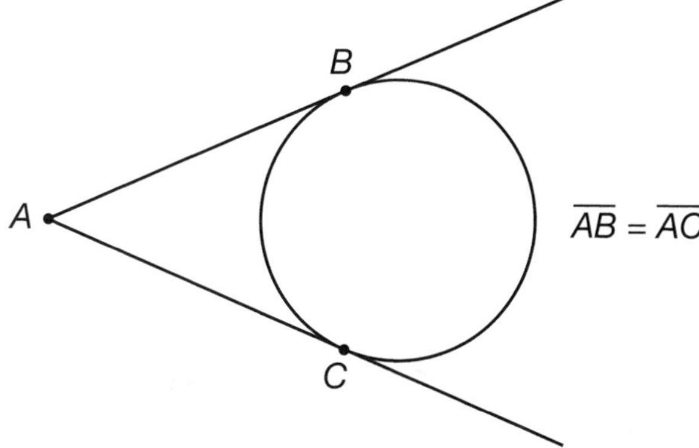

Practice Question

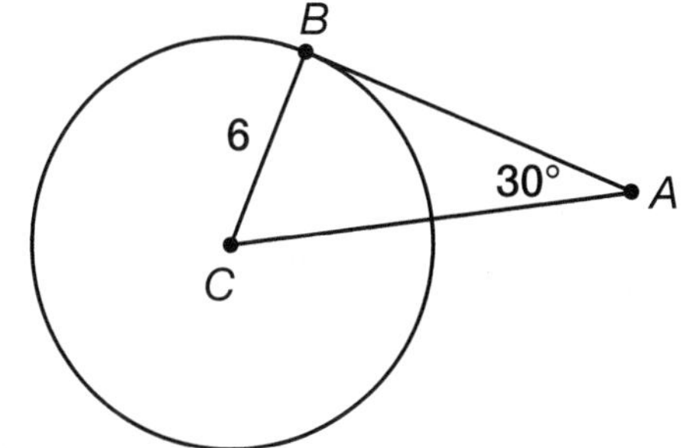

What is the length of $\overline{AB}$ in the figure above if $\overline{BC}$ is the radius of the circle and $\overline{AB}$ is tangent to the circle?

a. 3
b. $3\sqrt{2}$
c. $6\sqrt{2}$
d. $6\sqrt{3}$
e. 12

Answer

d. This problem requires knowledge of several rules of geometry. A tangent intersects with the radius of a circle at 90°. Therefore, ΔABC is a right triangle. Because one angle is 90° and another angle is 30°, then the third angle must be 60°. The triangle is therefore a 30-60-90 triangle.

In a 30-60-90 triangle, the leg opposite the 60° angle is $\sqrt{3} \times$ the leg opposite the 30° angle. In this figure, the leg opposite the 30° angle is 6, so $\overline{AB}$, which is the leg opposite the 60° angle, must be $6\sqrt{3}$.

▶ Polygons

A **polygon** is a closed figure with three or more sides.

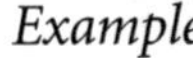

Example

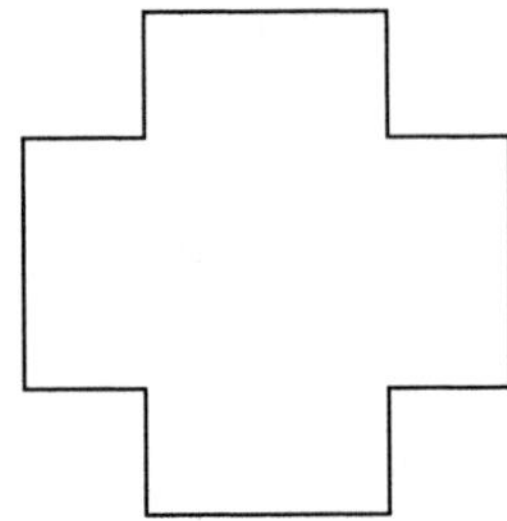

Terms Related to Polygons

- A **regular** (or equilateral) polygon has sides that are all equal; an **equiangular** polygon has angles that are all equal. The triangle below is a regular and equiangular polygon:

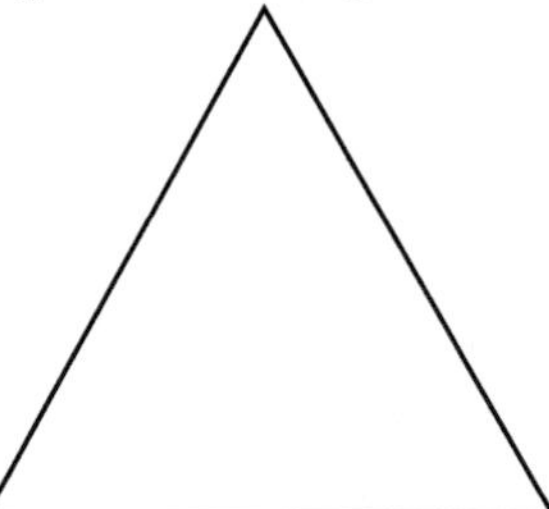

- Vertices are corner points of a polygon. The vertices in the six-sided polygon below are: *A*, *B*, *C*, *D*, *E*, and *F*.

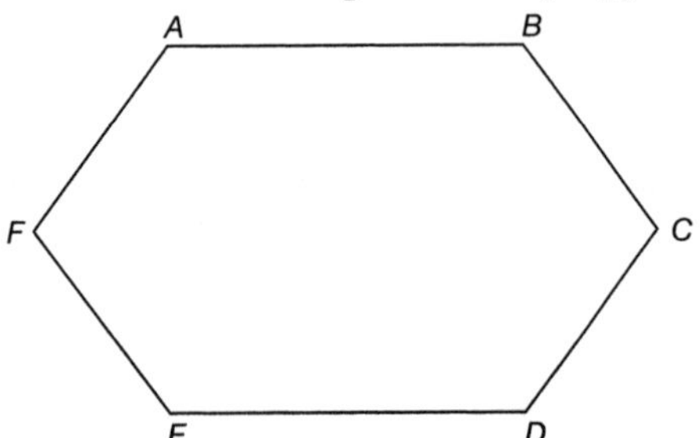

- A **diagonal** of a polygon is a line segment between two non-adjacent vertices. The diagonals in the polygon below are line segments $\overline{AC}$, $\overline{AD}$, $\overline{AE}$, $\overline{BD}$, $\overline{BE}$, $\overline{BF}$, $\overline{CE}$, $\overline{CF}$, and $\overline{DF}$.

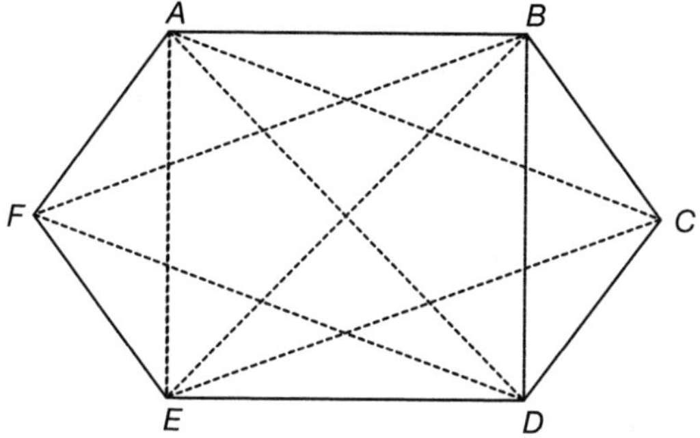

Quadrilaterals

A **quadrilateral** is a four-sided polygon. Any quadrilateral can be divided by a diagonal into two triangles, which means the sum of a quadrilateral's angles is $180° + 180° = 360°$.

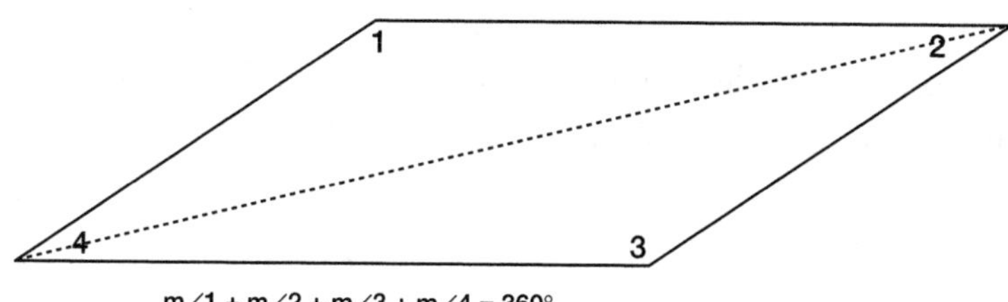

$m\angle 1 + m\angle 2 + m\angle 3 + m\angle 4 = 360°$

Sums of Interior and Exterior Angles

To find the sum of the **interior angles** of any polygon, use the following formula:

$S = 180(x - 2)$, with x being the number of sides in the polygon.

Example

Find the sum of the angles in the six-sided polygon below:

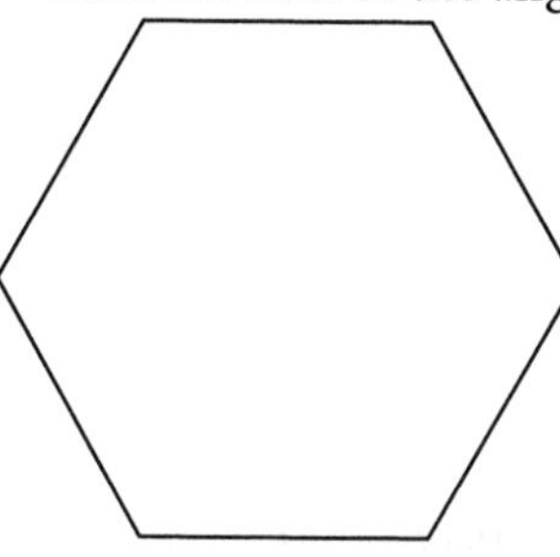

$S = 180(x - 2)$
$S = 180(6 - 2)$
$S = 180(4)$
$S = 720$
The sum of the angles in the polygon is 720°.

Practice Question

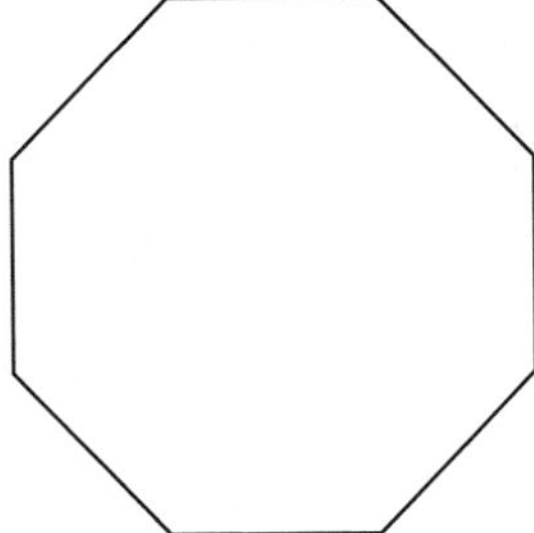

What is the sum of the interior angles in the figure above?

a. 360°
b. 540°
c. 900°
d. 1,080°
e. 1,260°

Answer

d. To find the sum of the interior angles of a polygon, use the formula $S = 180(x - 2)$, with x being the number of sides in the polygon. The polygon above has eight sides, therefore $x = 8$.
$S = 180(x - 2) = 180(8 - 2) = 180(6) = 1{,}080°$

Exterior Angles

The sum of the exterior angles of *any* polygon (triangles, quadrilaterals, pentagons, hexagons, etc.) is 360°.

Similar Polygons

If two polygons are similar, their corresponding angles are equal, and the ratio of the corresponding sides is in proportion.

Example

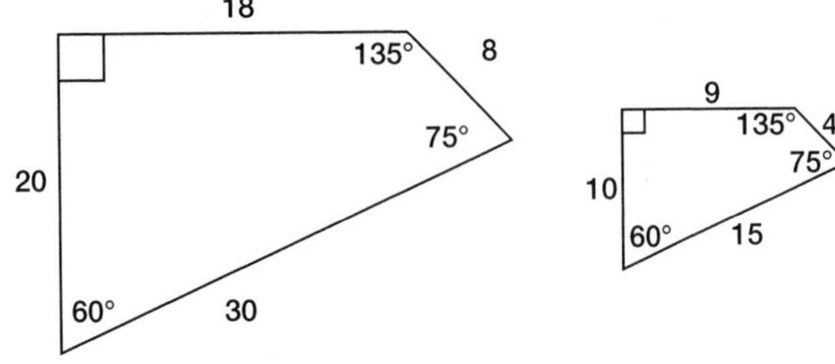

These two polygons are similar because their angles are equal and the ratio of the corresponding sides is in proportion:

$\frac{20}{10} = \frac{2}{1}$ $\frac{18}{9} = \frac{2}{1}$ $\frac{8}{4} = \frac{2}{1}$ $\frac{30}{15} = \frac{2}{1}$

Practice Question

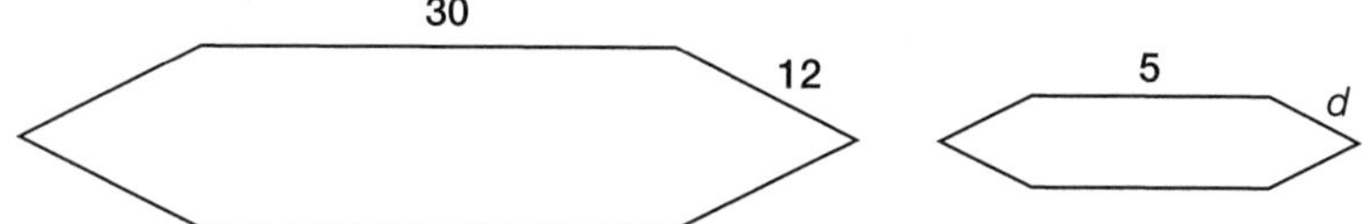

If the two polygons above are similar, what is the value of *d*?

a. 2
b. 5
c. 7
d. 12
e. 23

Answer

a. The two polygons are similar, which means the ratio of the corresponding sides are in proportion. Therefore, if the ratio of one side is 30:5, then the ration of the other side, 12:*d*, must be the same. Solve for *d* using proportions:

$\frac{30}{5} = \frac{12}{d}$ Find cross products.

$30d = (5)(12)$

$30d = 60$

$d = \frac{60}{30}$

$d = 2$

Parallelograms

A **parallelogram** is a quadrilateral with two pairs of parallel sides.

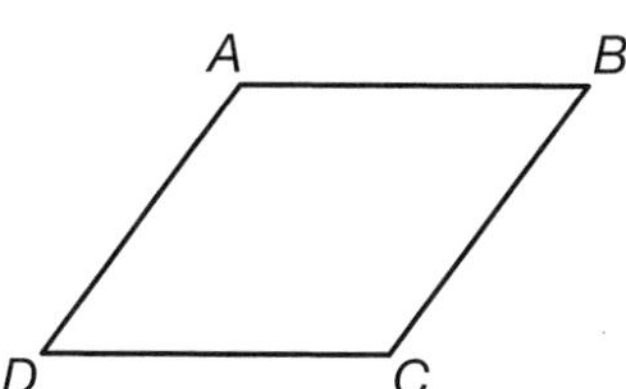

In the figure above, $\overline{AB} \parallel \overline{DC}$ and $\overline{AD} \parallel \overline{BC}$.

Parallelograms have the following attributes:

- opposite sides that are equal
  $\overline{AD} = \overline{BC}$ $\overline{AB} = \overline{DC}$
- opposite angles that are equal
  $m\angle A = m\angle C$ $m\angle B = m\angle D$
- consecutive angles that are supplementary
  $m\angle A + m\angle B = 180°$ $m\angle B + m\angle C = 180°$
  $m\angle C + m\angle D = 180°$ $m\angle D + m\angle A = 180°$

Special Types of Parallelograms

- A **rectangle** is a parallelogram with four right angles.

$\overline{AD} = \overline{BC}$ $\overline{AB} = \overline{DC}$

- A **rhombus** is a parallelogram with four equal sides.

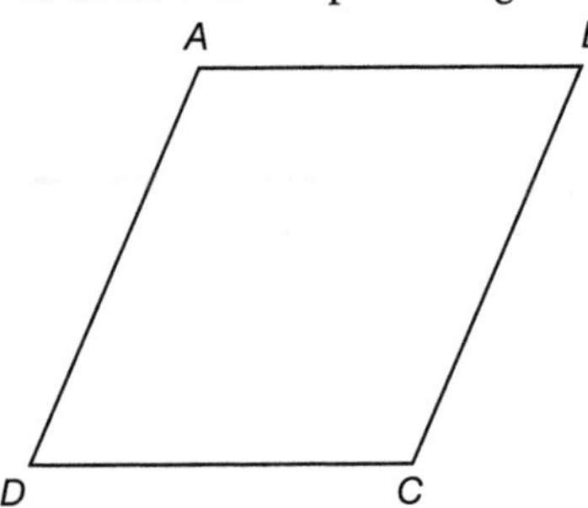

$\overline{AB} = \overline{BC} = \overline{DC} = \overline{AD}$

- A **square** is a parallelogram with four equal sides and four right angles.

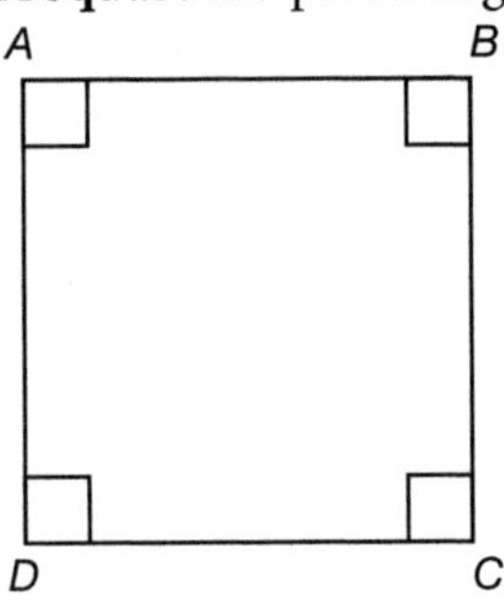

$\overline{AB} = \overline{BC} = \overline{DC} = \overline{AD}$

$m\angle A = m\angle B = m\angle C = m\angle D = 90$

Diagonals

- A diagonal cuts a parallelogram into two equal halves.

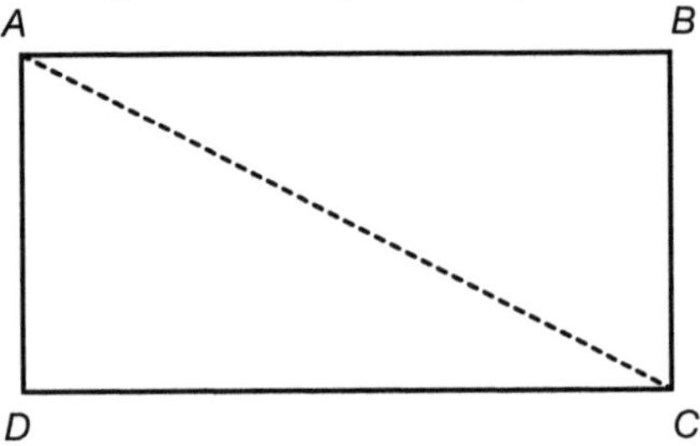

$\triangle ABC = \triangle ADC$

- In all parallelograms, diagonals cut each other into two equal halves.

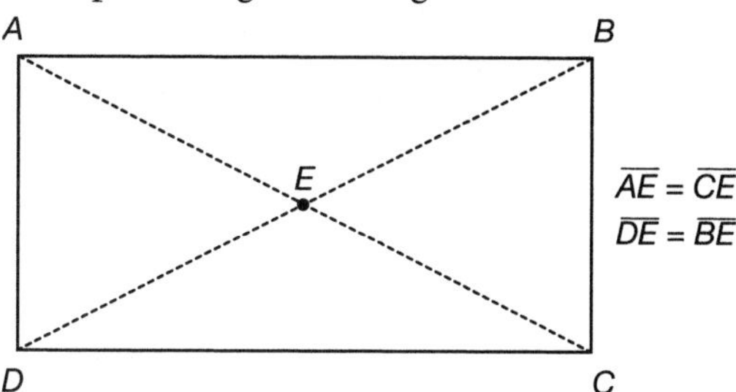

- In a rectangle, diagonals are the same length.

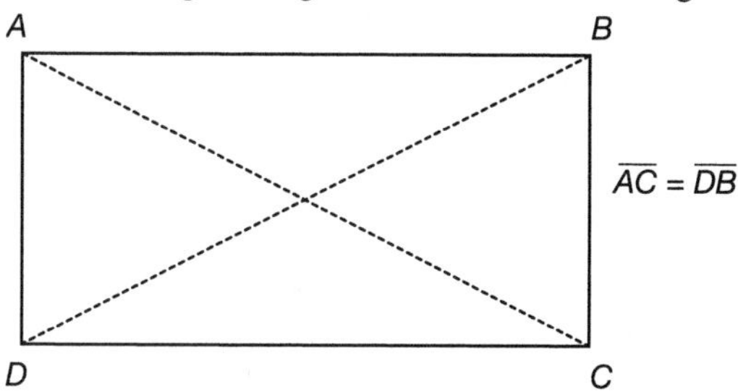

- In a rhombus, diagonals intersect at right angles.

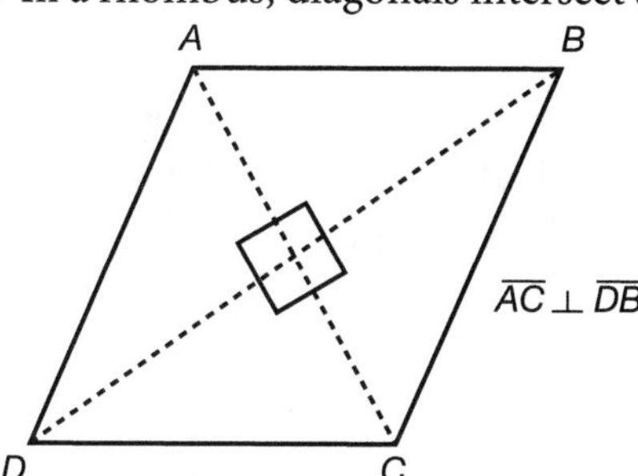

- In a square, diagonals are the same length and intersect at right angles.

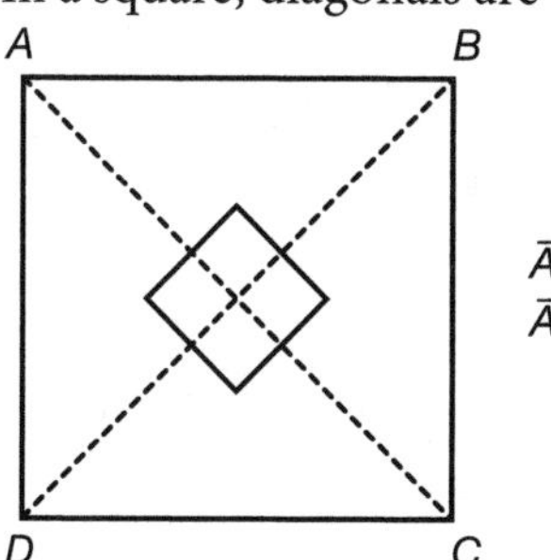

Practice Question

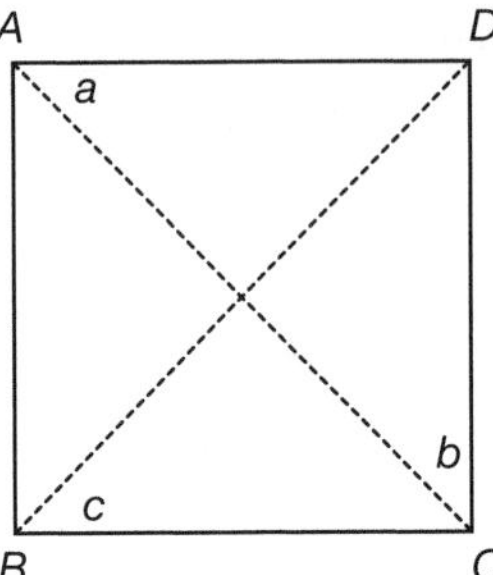

Which of the following must be true about the square above?

I. $a = b$

II. $\overline{AC} = \overline{BD}$

III. $b = c$

a. I only
b. II only
c. I and II only
d. II and III only
e. I, II, and III

Answer

e. $\overline{AC}$ and $\overline{BD}$ are diagonals. Diagonals cut parallelograms into two equal halves. Therefore, the diagonals divide the square into two 45-45-90 right triangles. Therefore, a, b, and c each equal 45°.

Now we can evaluate the three statements:

I: $a = b$ is TRUE because $a = 45$ and $b = 45$.

II: $\overline{AC} = \overline{BD}$ is TRUE because diagonals are equal in a square.

III: $b = c$ is TRUE because $b = 45$ and $c = 45$.

Therefore I, II, and III are ALL TRUE.

▶ Solid Figures, Perimeter, and Area

There are five kinds of measurement that you must understand for the SAT:

1. The **perimeter** of an object is the sum of all of its sides.

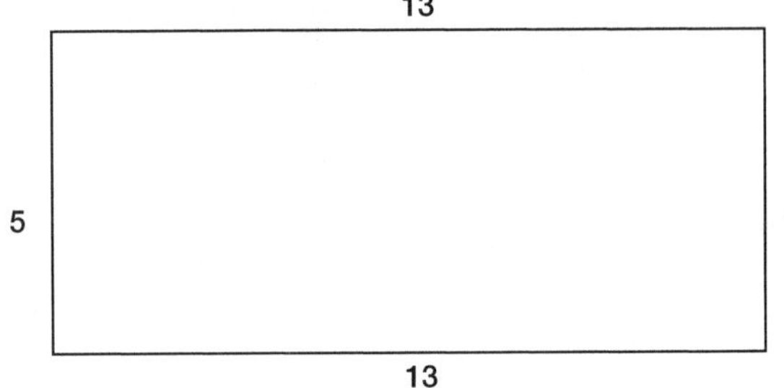

Perimeter $= 5 + 13 + 5 + 13 = 36$

2. **Area** is the number of square units that can fit inside a shape. Square units can be square inches (in^2), square feet (ft^2), square meters (m^2), etc.

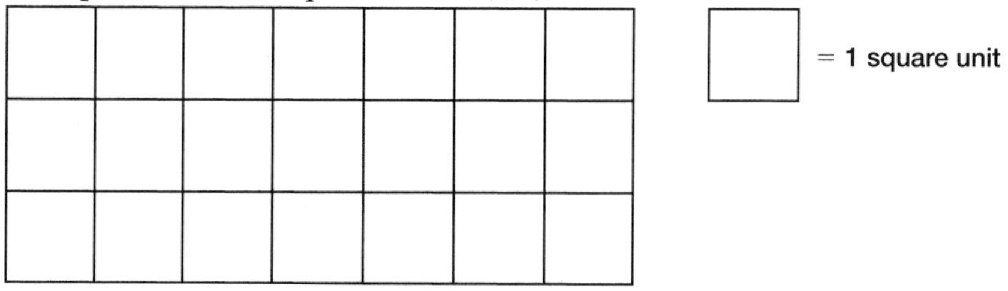

= 1 square unit

The area of the rectangle above is 21 square units. 21 square units fit inside the rectangle.

3. **Volume** is the number of cubic units that fit inside solid. Cubic units can be cubic inches (in^3), cubic feet (ft^2), cubic meters (m^3), etc.

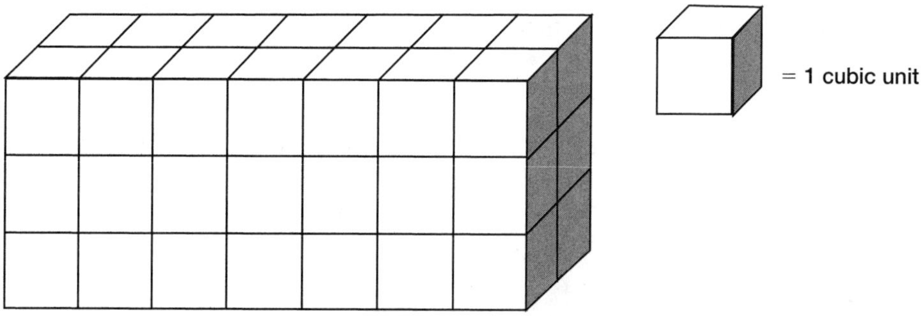

= 1 cubic unit

The volume of the solid above is 36 cubic units. 36 cubic units fit inside the solid.

4. The **surface area** of a solid is the sum of the areas of all its faces.

To find the surface area of this solid . . .

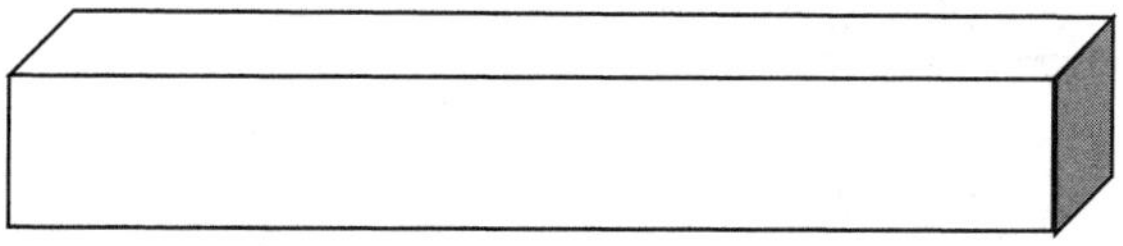

. . . add the areas of the four rectangles and the two squares that make up the surfaces of the solid.

5. Circumference is the distance around a circle.

If you uncurled this circle . . .

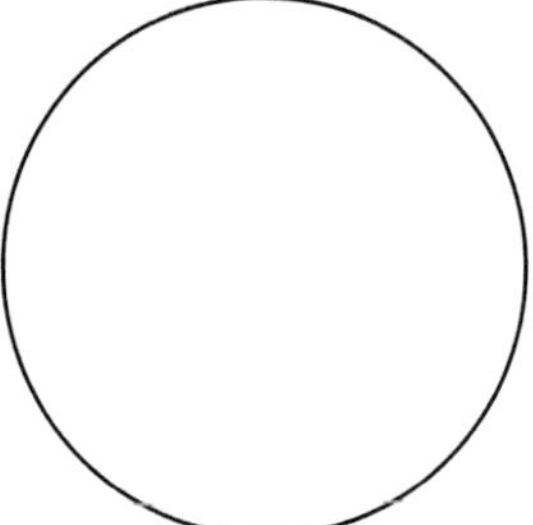

. . . you would have this line segment:

The circumference of the circle is the length of this line segment.

Formulas

The following formulas are provided on the SAT. You therefore do not need to memorize these formulas, but you do need to understand when and how to use them.

Circle

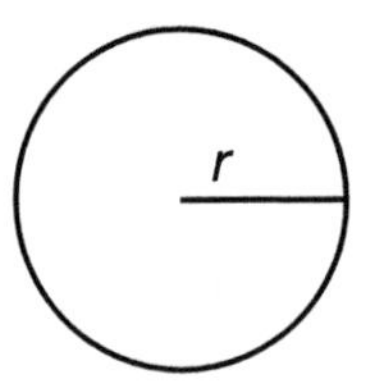

$C = 2\pi r$
$A = \pi r^2$

Rectangle

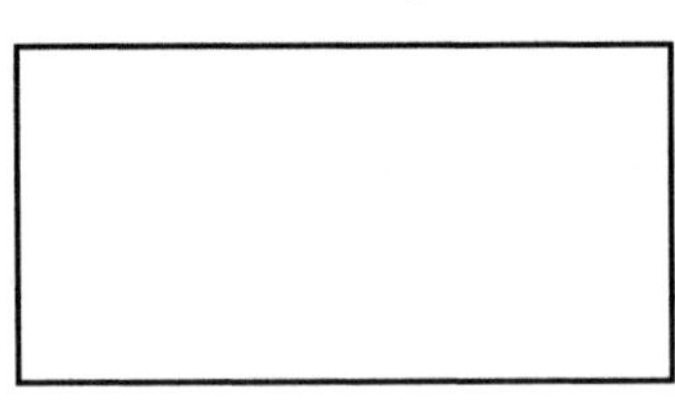

$A = lw$

Triangle

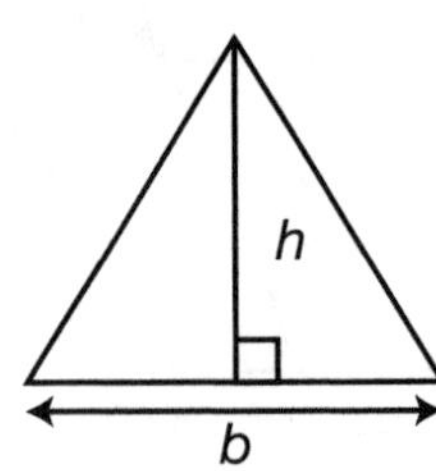

$A = \frac{1}{2}bh$

Cylinder

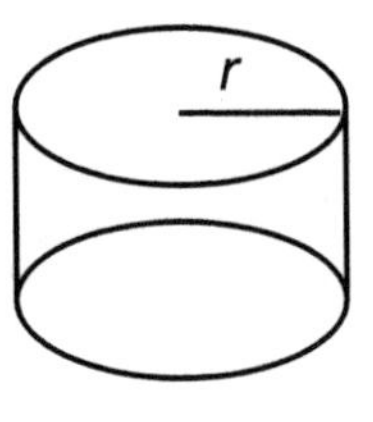

$V = \pi r^2 h$

Rectangle Solid

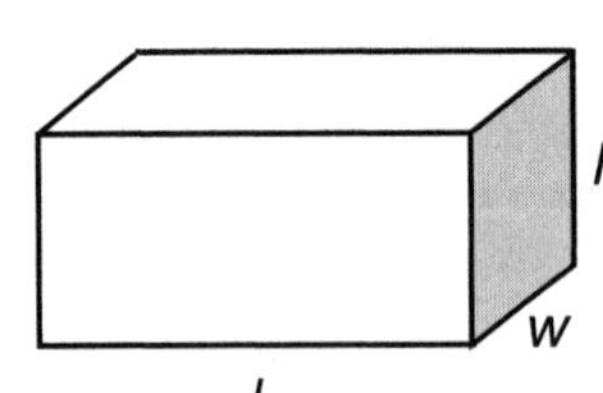

$V = lwh$

C =	Circumference	w =	Width
A =	Area	h =	Height
r =	Radius	V =	Volume
l =	Length	b =	Base

Practice Question

A rectangle has a perimeter of 42 and two sides of length 10. What is the length of the other two sides?

a. 10
b. 11
c. 22
d. 32
e. 52

Answer

b. You know that the rectangle has two sides of length 10. You also know that the other two sides of the rectangle are equal because rectangles have two sets of equal sides. Draw a picture to help you better understand:

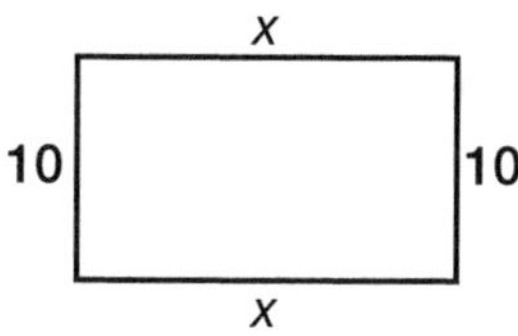

Based on the figure, you know that the perimeter is $10 + 10 + x + x$. So set up an equation and solve for x:

$10 + 10 + x + x = 42$

$20 + 2x = 42$

$20 + 2x - 20 = 42 - 20$

$2x = 22$

$\frac{2x}{2} = \frac{22}{2}$

$x = 11$

Therefore, we know that the length of the other two sides of the rectangle is 11.

Practice Question

The height of a triangular fence is 3 meters less than its base. The base of the fence is 7 meters. What is the area of the fence in square meters?

a. 4
b. 10
c. 14
d. 21
e. 28

Answer

c. Draw a picture to help you better understand the problem. The triangle has a base of 7 meters. The height is three meters less than the base ($7 - 3 = 4$), so the height is 4 meters:

4

7

The formula for the area of a triangle is $\frac{1}{2}$(base)(height):

$A = \frac{1}{2}bh$

$A = \frac{1}{2}(7)(4)$

$A = \frac{1}{2}(28)$

$A = 14$

The area of the triangular wall is 14 square meters.

Practice Question

A circular cylinder has a radius of 3 and a height of 5. Ms. Stewart wants to build a rectangular solid with a volume as close as possible to the cylinder. Which of the following rectangular solids has dimension closest to that of the circular cylinder?

a. $3 \times 3 \times 5$
b. $3 \times 5 \times 5$
c. $2 \times 5 \times 9$
d. $3 \times 5 \times 9$
e. $5 \times 5 \times 9$

Answer

d. First determine the approximate volume of the cylinder. The formula for the volume of a cylinder is $V = \pi r^2 h$. (Because the question requires only an approximation, use $\pi \approx 3$ to simplify your calculation.)

$V = \pi r^2 h$

$V \approx (3)(3^2)(5)$

$V \approx (3)(9)(5)$

$V \approx (27)(5)$

$V \approx 135$

Now determine the answer choice with dimensions that produce a volume closest to 135:

Answer choice **a:** $3 \times 3 \times 5 = 9 \times 5 = 45$

Answer choice **b:** $3 \times 5 \times 5 = 15 \times 5 = 75$

Answer choice **c:** $2 \times 5 \times 9 = 10 \times 9 = 90$

Answer choice **d:** $3 \times 5 \times 9 = 15 \times 9 = 135$

Answer choice **e:** $5 \times 5 \times 9 = 25 \times 9 = 225$

Answer choice **d** equals 135, which is the same as the approximate volume of the cylinder.

Practice Question

Mr. Suarez painted a circle with a radius of 6. Ms. Stone painted a circle with a radius of 12. How much greater is the circumference of Ms. Stone's circle than Mr. Suarez's circle?

a. 3π
b. 6π
c. 12π
d. 108π
e. 216π

Answer

c. You must determine the circumferences of the two circles and then subtract. The formula for the circumference of a circle is $C = 2\pi r$.

Mr. Suarez's circle has a radius of 6:
$C = 2\pi r$
$C = 2\pi(6)$
$C = 12\pi$

Ms. Stone's circle has a radius of 12:
$C = 2\pi r$
$C = 2\pi(12)$
$C = 24\pi$

Now subtract:
$24\pi - 12\pi = 12\pi$

The circumference of Ms. Stone's circle is 12π greater than Mr. Suarez's circle.

▶ Coordinate Geometry

A **coordinate plane** is a grid divided into four quadrants by both a horizontal x-axis and a vertical y-axis. **Coordinate points** can be located on the grid using **ordered pairs**. Ordered pairs are given in the form of (x,y). The x represents the location of the point on the horizontal x-axis, and the y represents the location of the point on the vertical y-axis. The x-axis and y-axis intersect at the **origin**, which is coordinate point (0,0).

Graphing Ordered Pairs

The **x-coordinate** is listed first in the ordered pair, and it tells you how many units to move to either the left or the right. If the x-coordinate is positive, move from the origin to the right. If the x-coordinate is negative, move from the origin to the left.

The ***y*-coordinate** is listed second and tells you how many units to move up or down. If the *y*-coordinate is positive, move up from the origin. If the *y*-coordinate is negative, move down from the origin.

Example

Graph the following points:

(0,0) (3,5) (3,−5) (−3,5) (−3,−5)

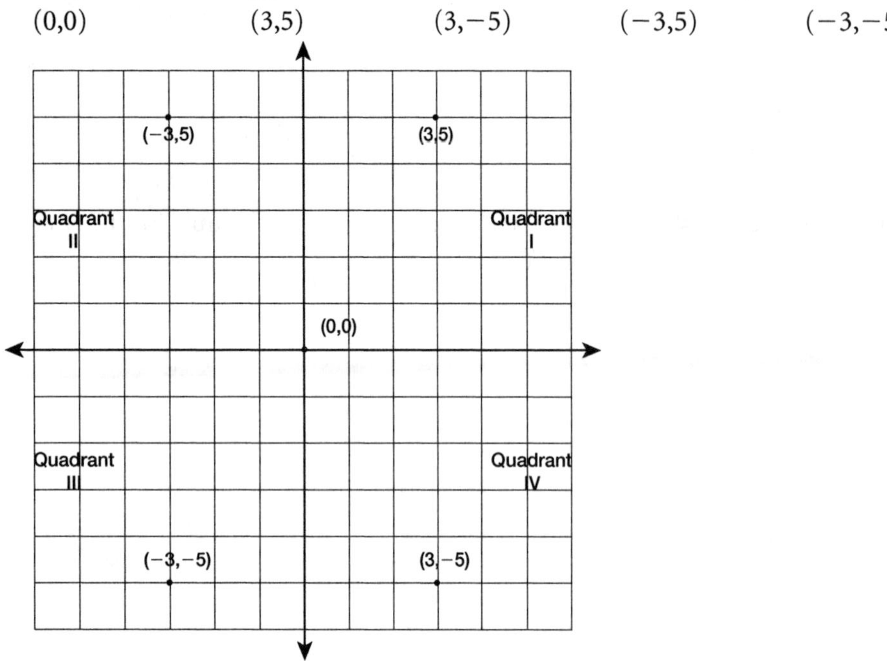

Notice that the graph is broken up into four quadrants with one point plotted in each one. The chart below indicates which quadrants contain which ordered pairs based on their signs:

POINT	SIGNS OF COORDINATES	QUADRANT
(3,5)	(+,+)	I
(–3,5)	(–,+)	II
(–3,–5)	(–,–)	III
(3,–5)	(+,–)	IV

Practice Question

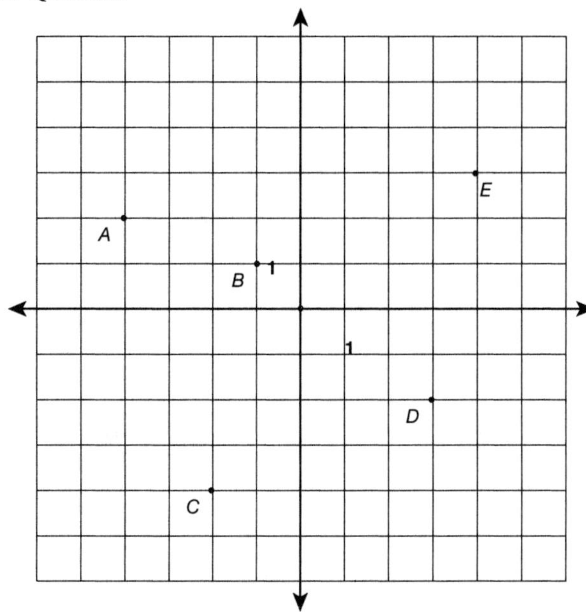

Which of the five points on the graph above has coordinates (x,y) such that $x + y = 1$?

a. *A*
b. *B*
c. *C*
d. *D*
e. *E*

Answer

d. You must determine the coordinates of each point and then add them:
A $(2,-4)$: $2 + (-4) = -2$
B $(-1,1)$: $-1 + 1 = 0$
C $(-2,-4)$: $-2 + (-4) = -6$
D $(3,-2)$: $3 + (-2) = 1$
E $(4,3)$: $4 + 3 = 7$
Point *D* is the point with coordinates (x,y) such that $x + y = 1$.

Lengths of Horizontal and Vertical Segments

The length of a horizontal or a vertical segment on the coordinate plane can be found by taking the absolute value of the difference between the two coordinates, which are different for the two points.

Example

Find the length of $\overline{AB}$ and $\overline{BC}$.

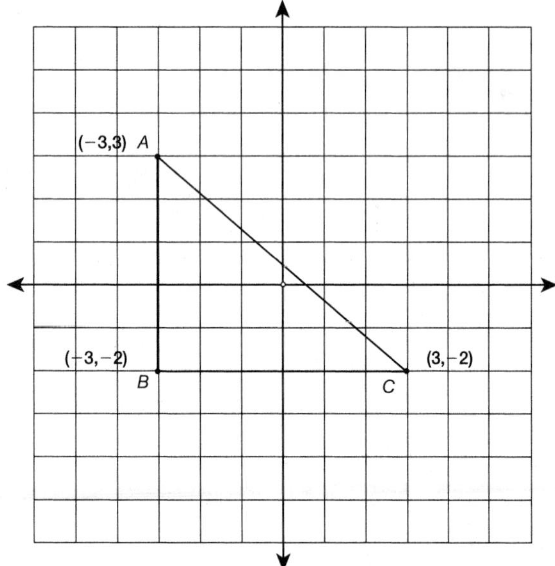

$\overline{AB}$ is parallel to the y-axis, so subtract the absolute value of the y-coordinates of its endpoints to find its length:

$\overline{AB} = |3 - (-2)|$

$\overline{AB} = |3 + 2|$

$\overline{AB} = |5|$

$\overline{AB} = 5$

$\overline{BC}$ is parallel to the x-axis, so subtract the absolute value of the x-coordinates of its endpoints to find its length:

$\overline{BC} = |-3 - 3|$

$\overline{BC} = |-6|$

$\overline{BC} = 6$

Practice Question

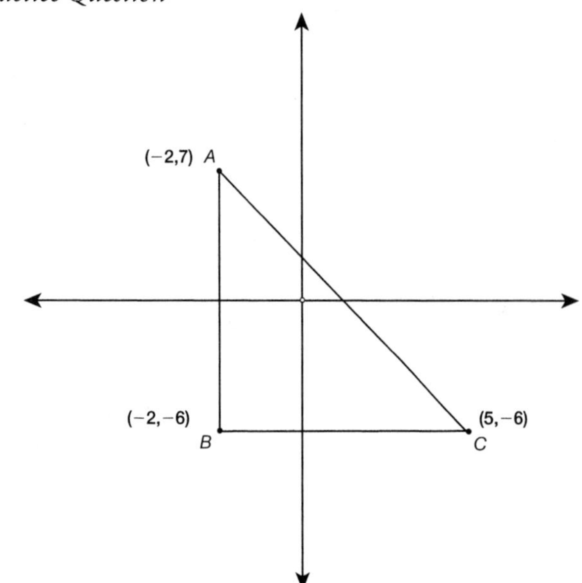

What is the sum of the length of $\overline{AB}$ and the length of $\overline{BC}$?

a. 6
b. 7
c. 13
d. 16
e. 20

Answer

e. $\overline{AB}$ is parallel to the y-axis, so subtract the absolute value of the y-coordinates of its endpoints to find its length:
$\overline{AB} = |7 - (-6)|$
$\overline{AB} = |7 + 6|$
$\overline{AB} = |13|$
$\overline{AB} = 13$
$\overline{BC}$ is parallel to the x-axis, so subtract the absolute value of the x-coordinates of its endpoints to find its length:
$\overline{BC} = |5 - (-2)|$
$\overline{BC} = |5 + 2|$
$\overline{BC} = |7|$
$\overline{BC} = 7$
Now add the two lengths: 7 + 13 = 20.

Distance between Coordinate Points

To find the distance between two points, use this variation of the Pythagorean theorem:

$$d = \sqrt{(x_2 - x_1)^2 + (y_2 - y_1)^2}$$

Example

Find the distance between points (2,−4) and (−3,−4).

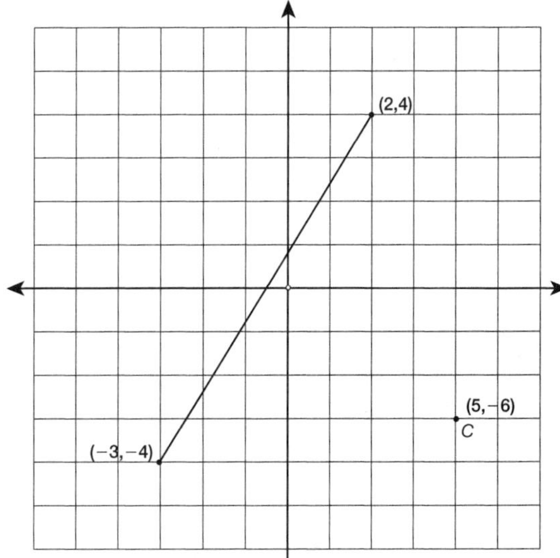

The two points in this problem are (2,−4) and (−3,−4).

$x_1 = 2$
$x_2 = -3$
$y_1 = -4$
$y_2 = -4$

Plug in the points into the formula:

$d = \sqrt{(x_2 - x_1)^2 + (y_2 - y_1)^2}$
$d = \sqrt{(-3 - 2)^2 + (-4 - (-4))^2}$
$d = \sqrt{(-3 - 2)^2 + (-4 + 4)^2}$
$d = \sqrt{(-5)^2 + (0)^2}$
$d = \sqrt{25}$
$d = 5$

The distance is 5.

Practice Question

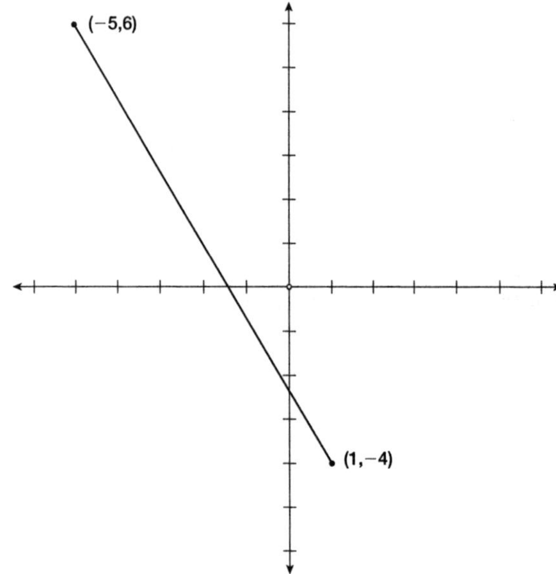

What is the distance between the two points shown in the figure above?

a. $\sqrt{20}$
b. 6
c. 10
d. $2\sqrt{34}$
e. $4\sqrt{34}$

Answer

d. To find the distance between two points, use the following formula:
$d = \sqrt{(x_2 - x_1)^2 + (y_2 - y_1)^2}$
The two points in this problem are $(-5,6)$ and $(1,-4)$.
$x_1 = -5$
$x_2 = 1$
$y_1 = 6$
$y_2 = -4$
Plug the points into the formula:
$d = \sqrt{(x_2 - x_1)^2 + (y_2 - y_1)^2}$
$d = \sqrt{(1 - (-5))^2 + (-4 - 6)^2}$
$d = \sqrt{(1 + 5)^2 + (-10)^2}$
$d = \sqrt{(6)^2 + (-10)^2}$
$d = \sqrt{36 + 100}$
$d = \sqrt{136}$
$d = \sqrt{4 \times 34}$
$d = \sqrt{34}$
The distance is $2\sqrt{34}$.

Midpoint

A **midpoint** is the point at the exact middle of a line segment. To find the midpoint of a segment on the coordinate plane, use the following formulas:

Midpoint $x = \frac{x_1 + x_2}{2}$ Midpoint $y = \frac{y_1 + y_2}{2}$

Example

Find the midpoint of $\overline{AB}$.

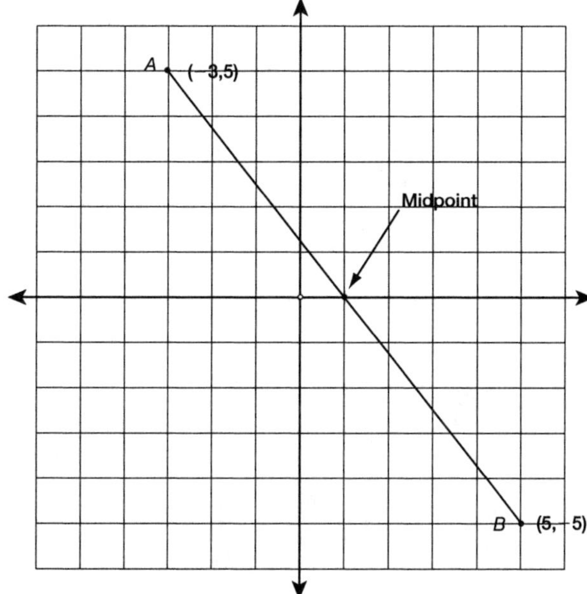

Midpoint $x = \frac{x_1 + x_2}{2} = \frac{-3 + 5}{2} = \frac{2}{2} = 1$

Midpoint $y = \frac{y_1 + y_2}{2} = \frac{5 + (-5)}{2} = \frac{0}{2} = 0$

Therefore, the midpoint of $\overline{AB}$ is (1,0).

Slope

The **slope** of a line measures its steepness. Slope is found by calculating the ratio of the change in y-coordinates of any two points on the line, over the change of the corresponding x-coordinates:

$$\text{slope} = \frac{\text{vertical change}}{\text{horizontal change}} = \frac{y_2 - y_1}{x_2 - x_1}$$

Example

Find the slope of a line containing the points (1,3) and (−3,−2).

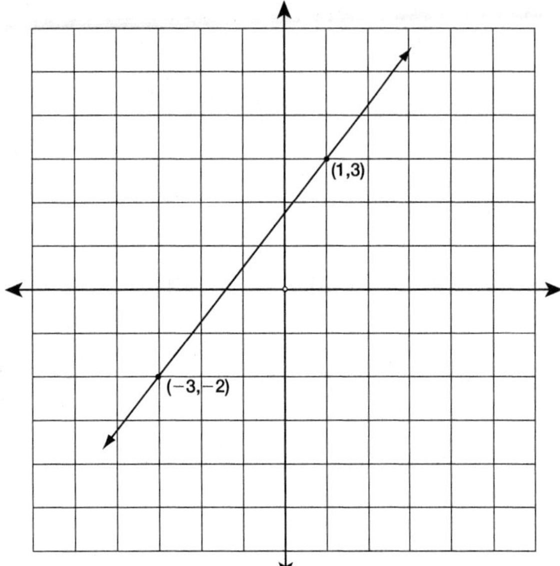

Slope $= \frac{y_2 - y_1}{x_2 - x_1} = \frac{3 - (-2)}{1 - (-3)} = \frac{3 + 2}{1 + 3} = \frac{5}{4}$

Therefore, the slope of the line is $\frac{5}{4}$.

Practice Question

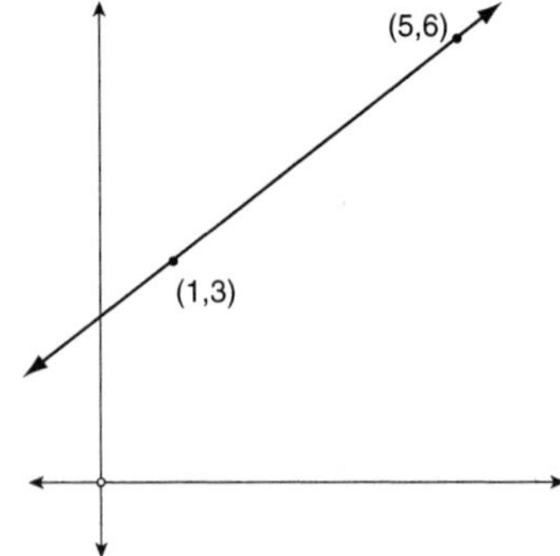

What is the slope of the line shown in the figure on the previous page?

a. $\frac{1}{2}$

b. $\frac{3}{4}$

c. $\frac{4}{3}$

d. 2

e. 3

Answer

b. To find the slope of a line, use the following formula:

$\text{slope} = \frac{\text{vertical change}}{\text{horizontal change}} = \frac{y_2 - y_1}{x_2 - x_1}$

The two points shown on the line are (1,3) and (5,6).

$x_1 = 1$

$x_2 = 5$

$y_1 = 3$

$y_2 = 6$

Plug in the points into the formula:

$\text{slope} = \frac{6 - 3}{5 - 2}$

$\text{slope} = \frac{3}{4}$

Using Slope

If you know the slope of a line and one point on the line, you can determine other coordinate points on the line. Because slope tells you the ratio of $\frac{\text{vertical change}}{\text{horizontal change}}$, you can simply move from the coordinate point you know the required number of units determined by the slope.

Example

A line has a slope of $\frac{6}{5}$ and passes through point (3,4). What is another point the line passes through?

The slope is $\frac{6}{5}$, so you know there is a vertical change of 6 and a horizontal change of 5. So, starting at point (3,4), add 6 to the y-coordinate and add 5 to the x-coordinate:

y: $4 + 6 = 10$

x: $3 + 5 = 8$

Therefore, another coordinate point is (8,10).

If you know the slope of a line and one point on the line, you can also determine a point at a certain coordinate, such as the y-intercept $(x,0)$ or the x-intercept $(0,y)$.

Example

A line has a slope of $\frac{2}{3}$ and passes through point (1,4). What is the y-intercept of the line?

Slope $= \frac{y_2 - y_1}{x_2 - x_1}$, so you can plug in the coordinates of the known point (1,4) and the unknown point, the y-intercept $(x,0)$, and set up a ratio with the known slope, $\frac{2}{3}$, and solve for x:

$\frac{y_2 - y_1}{x_2 - x_1} = \frac{2}{3}$

$\frac{0 - 4}{x - 1} = \frac{2}{3}$

$\frac{0-4}{x-1} = \frac{2}{3}$ Find cross products.

$(-4)(3) = 2(x-1)$

$-12 = 2x - 2$

$-12 + 2 = 2x - 2 + 2$

$-\frac{10}{2} = \frac{2x}{2}$

$-\frac{10}{2} = x$

$-5 = x$

Therefore, the x-coordinate of the y-intercept is -5, so the y-intercept is $(-5,0)$.

Facts about Slope

- A line that *rises to the right* has a positive slope.

- A line that *falls to the right* has a negative slope.

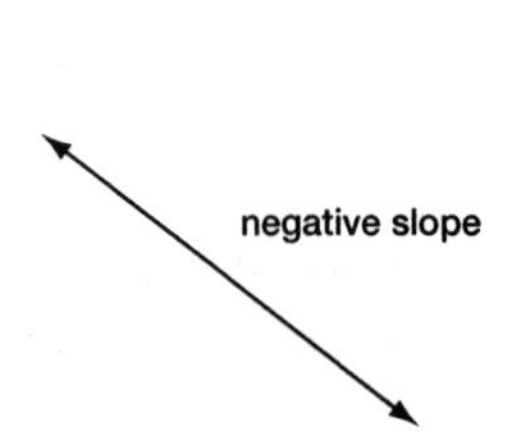

- A horizontal line has a slope of 0.

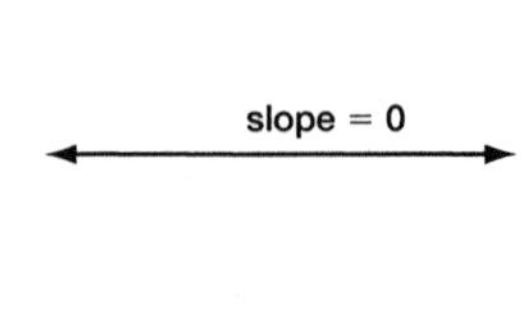

- A vertical line does not have a slope at all—it is undefined.

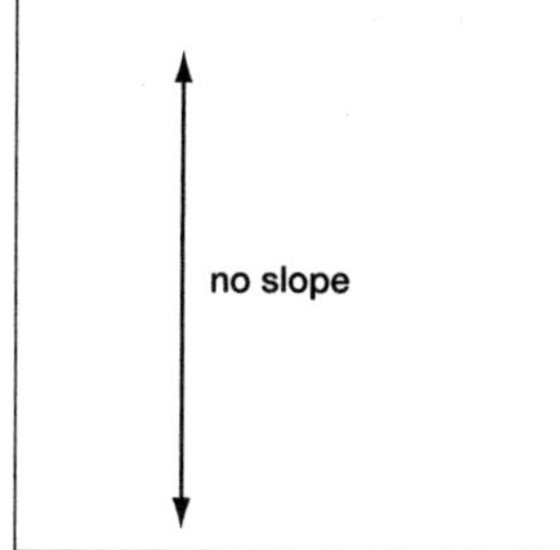

- Parallel lines have equal slopes.

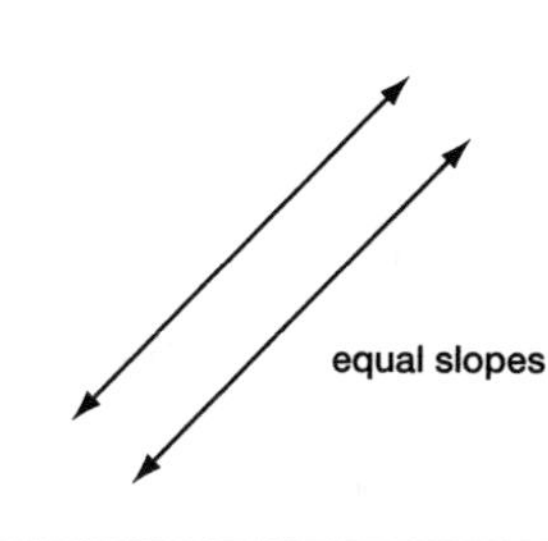

- Perpendicular lines have slopes that are negative reciprocals of each other (e.g., 2 and $-\frac{1}{2}$).

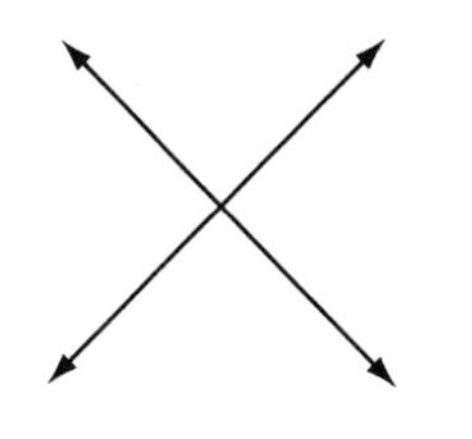

Practice Question

A line has a slope of -3 and passes through point (6,3). What is the y-intercept of the line?

a. (7,0)
b. (0,7)
c. (7,7)
d. (2,0)
e. (15,0)

Answer

a. Slope $= \frac{y_2 - y_1}{x_2 - x_1}$, so you can plug in the coordinates of the known point (6,3) and the unknown point, the y-intercept $(x,0)$, and set up a ratio with the known slope, -3, and solve for x:

$\frac{y_2 - y_1}{x_2 - x_1} = -3$

$\frac{0 - 3}{x - 6} = -3$

$\frac{-3}{x - 6} = -3$ Simplify.

$(x - 6)\frac{-3}{x - 6} = -3(x - 6)$

$-3 = -3x + 18$

$-3 - 18 = -3x + 18 - 18$

$-21 = -3x$

$\frac{-21}{-3} = \frac{-3x}{-3}$

$\frac{-21}{-3} = x$

$7 = x$

Therefore, the x-coordinate of the y-intercept is 7, so the y-intercept is (7,0).

CHAPTER

Problem Solving

This chapter reviews key problem-solving skills and concepts that you need to know for the SAT. Throughout the chapter are sample questions in the style of SAT questions. Each sample SAT question is followed by an explanation of the correct answer.

▶ Translating Words into Numbers

To solve word problems, you must be able to translate words into mathematical operations. You must analyze the language of the question and determine what the question is asking you to do.

The following list presents phrases commonly found in word problems along with their mathematical equivalents:

- A **number** means a **variable.**

Example

17 minus a number equals 4.

$17 - x = 4$

- **Increase** means **add**.

Example

a number increased by 8

$x + 8$

- **More than** means **add.**

Example

4 more than a number

$4 + x$

- **Less than** means **subtract.**

Example

8 less than a number

$x - 8$

- **Times** means **multiply.**

Example

6 times a number

$6x$

- **Times the sum** means to **multiply a number by a quantity.**

Example

7 times the sum of a number and 2

$7(x + 2)$

- Note that variables can be used together.

Example

A number y exceeds 3 times a number x by 12.

$y = 3x + 12$

- **Greater than** means $>$ and **less than** means $<$.

Examples

The product of x and 9 is greater than 15.

$x \times 9 > 15$

When 1 is added to a number x, the sum is less than 29.

$x + 1 < 29$

- **At least** means $\geq$ and **at most** means $\leq$.

Examples

The sum of a number x and 5 is at least 11.

$x + 5 \geq 11$

When 14 is subtracted from a number x, the difference is at most 6.

$x - 14 \leq 6$

- To **square** means to **use an exponent of 2.**

Example

The square of the sum of m and n is 25.

$(m + n)^2 = 25$

Practice Question

If squaring the sum of y and 23 gives a result that is 4 less than 5 times y, which of the following equations could you use to find the possible values of y?

a. $(y + 23)^2 = 5y - 4$
b. $y^2 + 23 = 5y - 4$
c. $y^2 + (23)^2 = y(4 - 5)$
d. $y^2 + (23)^2 = 5y - 4$
e. $(y + 23)^2 = y(4 - 5)$

Answer

a. Break the problem into pieces while translating into mathematics:
squaring translates to *raise something to a power of 2*
the sum of y *and 23* translates to $(y + 23)$
So, *squaring the sum of* y *and 23* translates to $(y + 23)^2$.
gives a result translates to =
4 less than translates to *something* − 4
5 times y translates to 5y
So, *4 less than 5 times* y means 5y − 4.
Therefore, *squaring the sum of* y *and 23 gives a result that is 4 less than 5 times* y translates to: $(y + 23)^2 = 5y - 4$.

▶ Assigning Variables in Word Problems

Some word problems require you to create and assign one or more variables. To answer these word problems, first identify the *unknown* numbers and the *known* numbers. Keep in mind that sometimes the "known" numbers won't be actual numbers, but will instead be expressions involving an unknown.

Examples

Renee is five years older than Ana.
Unknown = Ana's age = x
Known = Renee's age is five years more than Ana's age = $x + 5$
Paco made three times as many pancakes as Vince.
Unknown = number of pancakes Vince made = x
Known = number of pancakes Paco made = three times as many pancakes as Vince made = $3x$
Ahmed has four more than six times the number of CDs that Frances has.
Unknown = the number of CDs Frances has = x
Known = the number of CDs Ahmed has = four more than six times the number of CDs that Frances has = $6x + 4$

Practice Question

On Sunday, Vin's Fruit Stand had a certain amount of apples to sell during the week. On each subsequent day, Vin's Fruit Stand had one-fifth the amount of apples than on the previous day. On Wednesday, 3 days later, Vin's Fruit Stand had 10 apples left. How many apples did Vin's Fruit Stand have on Sunday?

a. 10
b. 50
c. 250
d. 1,250
e. 6,250

Answer

d. To solve, make a list of the knowns and unknowns:

Unknown:

Number of apples on **Sunday** = x

Knowns:

Number of apples on **Monday** = one-fifth the number of apples on Sunday = $\frac{1}{5}x$
Number of apples on **Tuesday** = one-fifth the number of apples on Monday = $\frac{1}{5}(\frac{1}{5}x)$
Number of apples on **Wednesday** = one-fifth the number of apples on Tuesday = $\frac{1}{5}[\frac{1}{5}(\frac{1}{5}x)]$
Because you know that Vin's Fruit Stand had 10 apples on Wednesday, you can set the expression for the number of apples on Wednesday equal to 10 and solve for x:

$\frac{1}{5}[\frac{1}{5}(\frac{1}{5}x)] = 10$

$\frac{1}{5}[\frac{1}{25}x] = 10$

$\frac{1}{125}x = 10$

$125 \times \frac{1}{125}x = 125 \times 10$

$x = 1{,}250$

Because x = the number of apples on Sunday, you know that Vin's Fruit Stand had 1,250 apples on Sunday.

▶ Percentage Problems

There are three types of percentage questions you might see on the SAT:

1. finding the percentage of a given number
   Example: What number is 60% of 24?
2. finding a number when a percentage is given
   Example: 30% of what number is 15?
3. finding what percentage one number is of another number
   Example: What percentage of 45 is 5?

To answer percent questions, write them as fraction problems. To do this, you must translate the questions into math. Percent questions typically contain the following elements:

- The **percent** is a number divided by 100.
  $75\% = \frac{75}{100} = 0.75$ $\quad 4\% = \frac{4}{100} = 0.04$ $\quad 0.3\% = \frac{0.3}{100} = 0.003$
- The word **of** means to multiply.
  English: 10% **of** 30 equals 3.
  Math: $\frac{10}{100} \times 30 = 3$
- The word **what** refers to a variable.
  English: 20% of **what** equals 8?
  Math: $\frac{20}{100} \times a = 8$
- The words **is**, **are**, and **were**, mean equals.
  English: 0.5% of 18 **is** 0.09.
  Math: $\frac{0.05}{100} \times 18 = 0.09$

When answering a percentage problem, rewrite the problem as math using the translations above and then solve.

- finding the percentage of a given number

Example

What number is 80% of 40?

First translate the problem into math:

What number is 80% of 40?

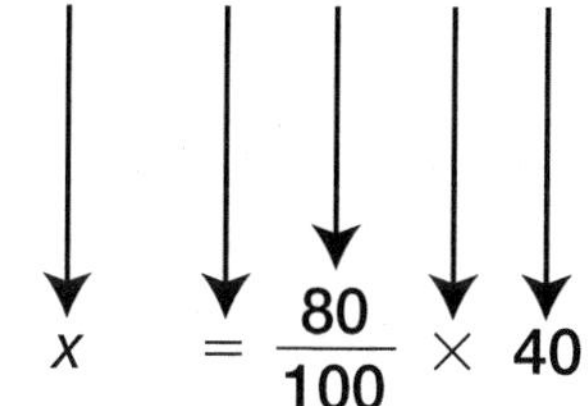

Now solve:

$x = \frac{80}{100} \times 40$

$x = \frac{3{,}200}{100}$

$x = 32$

Answer: 32 is 80% of 40

- finding a number that is a percentage of another number

Example

25% of what number is 16?

First translate the problem into math:

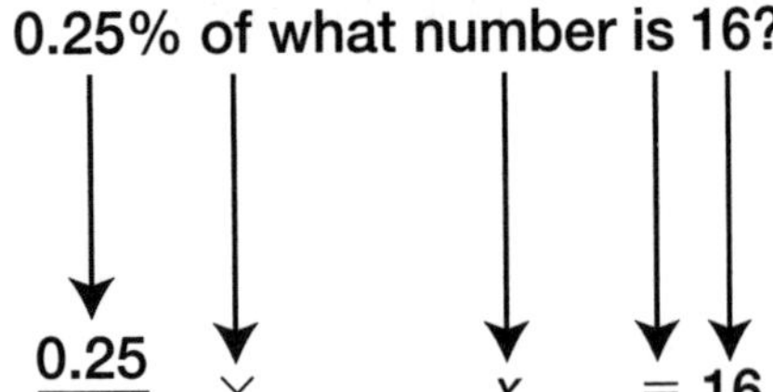

Now solve:

$\frac{0.25}{100} \times x = 16$

$\frac{0.25x}{100} = 16$

$\frac{0.25x}{100} \times 100 = 16 \times 100$

$0.25x = 1{,}600$

$\frac{x}{0.25} = \frac{1{,}600}{0.25}$

$x = 6{,}400$

Answer: 0.25% of 6,400 is 16.

- finding what percentage one number is of another number

Example

What percentage of 90 is 18?

First translate the problem into math:

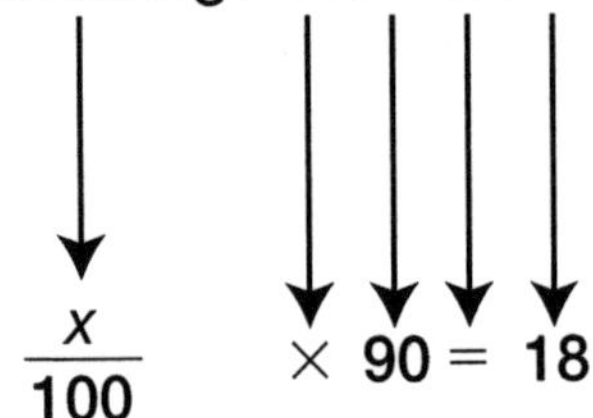

Now solve:

$\frac{x}{100} \times 90 = 18$

$\frac{90x}{100} = 18$

$\frac{9x}{10} = 18$

$\frac{9x}{10} \times 10 = 18 \times 10$

$9x = 180$

$x = 20$

Answer: 18 is 20% of 90.

Practice Question

If z is 2% of 85, what is 2% of z?

a. 0.034
b. 0.34
c. 1.7
d. 3.4
e. 17

Answer

a. To solve, break the problem into pieces. The first part says that z is 2% of 85. Let's translate:

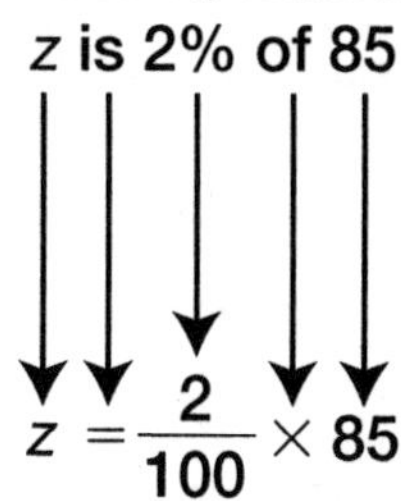

Now let's solve for z:

$z = \frac{2}{100} \times 85$

$z = \frac{1}{50} \times 85$

$z = \frac{85}{50}$

$z = \frac{17}{10}$

Now we know that $z = \frac{17}{10}$. The second part asks: What is 2% of z? Let's translate:

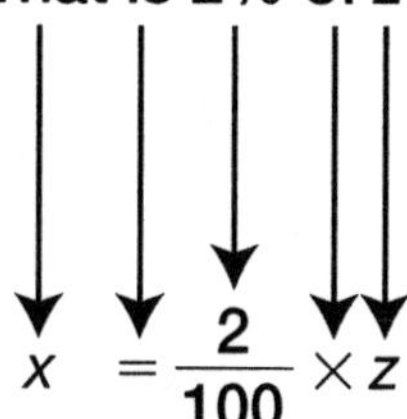

Now let's solve for x when $z = \frac{17}{10}$.

$x = \frac{2}{100} \times z$ Plug in the value of z.

$x = \frac{2}{100} \times \frac{17}{10}$

$x = \frac{34}{1{,}000} = 0.034$

Therefore, 0.034 is 2% of z.

▶ Ratios

A **ratio** is a comparison of two quantities measured in the same units. Ratios are represented with a colon or as a fraction:

$x{:}y$ $\frac{x}{y}$

$3{:}2$ $\frac{3}{2}$

$a{:}9$ $\frac{a}{9}$

Examples

If a store sells apples and oranges at a ratio of 2:5, it means that for every two apples the store sells, it sells 5 oranges.

If the ratio of boys to girls in a school is 13:15, it means that for every 13 boys, there are 15 girls.

Ratio problems may ask you to determine the number of items in a group based on a ratio. You can use the concept of multiples to solve these problems.

Example

A box contains 90 buttons, some blue and some white. The ratio of the number of blue to white buttons is 12:6. How many of each color button is in the box?

We know there is a ratio of 12 blue buttons to every 6 white buttons. This means that for every batch of 12 buttons in the box there is also a batch of 6 buttons. We also know there is a total of 90 buttons. This means that we must determine how many batches of blue and white buttons add up to a total of 90. So let's write an equation:

$12x + 6x = 90$, where x is the number of batches of buttons

$18x = 90$

$x = 5$

So we know that there are 5 batches of buttons.

Therefore, there are $(5 \times 12) = 60$ blue buttons and $(5 \times 6) = 30$ white buttons.

A **proportion** is an equality of two ratios.

$\frac{x}{6} = \frac{4}{7}$ $\frac{1}{35} = \frac{2}{a}$

You can use proportions to solve ratio problems that ask you to determine how much of something is needed based on how much you have of something else.

Example

A recipe calls for peanuts and raisins in a ratio of 3:4, respectively. If Carlos wants to make the recipe with 9 cups of peanuts, how many cups of raisins should he use?

Let's set up a proportion to determine how many cups of raisins Carlos needs.

$\frac{3}{4} = \frac{9}{r}$

This proportion means that 3 parts peanuts to 4 parts raisins must equal 9 parts peanuts to r parts raisins. We can solve for r by finding cross products:

$\frac{3}{4} = \frac{9}{r}$

$3r = 4 \times 9$

$3r = 36$

$\frac{3r}{3} = \frac{36}{3}$

$r = 12$

Therefore, if Carlos uses 9 cups of peanuts, he needs to use 12 cups of raisins.

Practice Question

A painter mixes red, green, and yellow paint in the ratio of 6:4:2 to produce a new color. In order to make 6 gallons of this new color, how many gallons of red paint must the painter use?

a. 1
b. 2
c. 3
d. 4
e. 6

Answer

c. In the ratio 6:4:2, we know there are 6 parts red paint, 4 parts green paint, and 2 parts yellow paint. Now we must first determine how many total parts there are in the ratio:

6 parts red + 4 parts green + 2 parts yellow = 12 total parts

This means that for every 12 parts of paint, 6 parts are red, 4 parts are green, and 2 parts are yellow. We can now set up a new ratio for red paint:

6 parts red paint:12 total parts = 6:12 = $\frac{6}{12}$

Because we need to find how many gallons of red paint are needed to make 6 total gallons of the new color, we can set up an equation to determine how many parts of red paint are needed to make 6 total parts:

$\frac{r\text{ parts red paint}}{6\text{ parts total}} = \frac{6\text{ parts red paint}}{12\text{ parts total}}$

$\frac{r}{6} = \frac{6}{12}$

Now let's solve for r:

$\frac{r}{6} = \frac{6}{12}$ Find cross products.

$12r = 6 \times 6$

$\frac{12r}{12} = \frac{36}{12}$

$r = 3$

Therefore, we know that 3 parts red paint are needed to make 6 total parts of the new color. So 3 gallons of red paint are needed to make 6 gallons of the new color.

▶ Variation

Variation is a term referring to a constant ratio in the change of a quantity.

- A quantity is said to **vary directly** with or to be **directly proportional to** another quantity if they both change in an equal direction. In other words, two quantities vary directly if an increase in one causes an increase in the other or if a decrease in one causes a decrease in the other. The ratio of increase or decrease, however, must be the same.

Example

Thirty elephants drink altogether a total of 6,750 liters of water a day. Assuming each elephant drinks the same amount, how many liters of water would 70 elephants drink?

Since each elephant drinks the same amount of water, you know that elephants and water vary directly. Therefore, you can set up a proportion:

$\frac{\text{water}}{\text{elephants}} = \frac{6{,}750}{30} = \frac{x}{70}$

Find cross products to solve:

$\frac{6{,}750}{30} = \frac{x}{70}$

$(6{,}750)(70) = 30x$

$472{,}500 = 30x$

$\frac{472{,}500}{30} = \frac{30x}{30}$

$15{,}750 = x$

Therefore, 70 elephants would drink 15,750 liters of water.

- A quantity is said to **vary inversely** with or to be **inversely proportional** to another quantity if they change in opposite directions. In other words, two quantities vary inversely if an increase in one causes a decrease in the other or if a decrease in one causes an increase in the other.

Example

Three plumbers can install plumbing in a house in six days. Assuming each plumber works at the same rate, how many days would it take nine plumbers to install plumbing in the same house?

As the number of plumbers increases, the days needed to install plumbing decreases (because more plumbers can do more work). Therefore, the relationship between the number of plumbers and the number of days varies inversely. Because the amount of plumbing to install remains constant, the two expressions can be set equal to each other:

3 plumbers $\times$ 6 days = 9 plumbers $\times$ x days

$3 \times 6 = 9x$

$18 = 9x$

$\frac{18}{9} = \frac{9x}{9}$

$2 = x$

Thus, it would take nine plumbers only two days to install plumbing in the same house.

Practice Question

The number a is directly proportional to b. If $a = 15$ when $b = 24$, what is the value of b when $a = 5$?

a. $\frac{8}{5}$

b. $\frac{25}{8}$

c. 8

d. 14

e. 72

Answer

c. The numbers a and b are directly proportional (in other words, they vary directly), so a increases when b increases, and vice versa. Therefore, we can set up a proportion to solve:

$\frac{15}{24} = \frac{5}{b}$ Find cross products.

$15b = (24)(5)$

$15b = 120$

$\frac{15b}{15} = \frac{120}{15}$

$b = 8$

Therefore, we know that $b = 8$ when $a = 5$.

▶ Rate Problems

Rate is defined as a comparison of two quantities with different units of measure.

$$\text{Rate} = \frac{x \text{ units}}{y \text{ units}}$$

Examples

$\frac{\text{dollars}}{\text{hour}}$ $\frac{\text{cost}}{\text{pound}}$ $\frac{\text{miles}}{\text{hour}}$ $\frac{\text{miles}}{\text{gallon}}$

There are three types of rate problems you must learn how to solve: cost per unit problems, movement problems, and work-output problems.

▶ Cost Per Unit

Some rate problems require you to calculate the cost of a specific quantity of items.

Example

If 40 sandwiches cost \$298, what is the cost of eight sandwiches?

First determine the cost of one sandwich by setting up a proportion:

$\frac{\$238}{40 \text{ sandwiches}} = \frac{x}{1}$ sandwich

$238 \times 1 = 40x$ Find cross products.

$238 = 40x$

$\frac{238}{40} = x$

$5.95 = x$

Now we know one sandwich costs \$5.95. To find the cost of eight sandwiches, multiply:

$5.95 \times 8 = \$47.60$

Eight sandwiches cost \$47.60.

Practice Question

A clothing store sold 45 bandanas a day for three days in a row. If the store earned a total of \$303.75 from the bandanas for the three days, and each bandana cost the same amount, how much did each bandana cost?

a. \$2.25
b. \$2.75
c. \$5.50
d. \$6.75
e. \$101.25

Answer

a. First determine how many total bandanas were sold:

45 bandanas per day $\times$ 3 days = 135 bandanas

So you know that 135 bandanas cost \$303.75. Now set up a proportion to determine the cost of one bandana:

$\frac{\$303.75}{135 \text{ bandanas}} = \frac{x}{1 \text{ bandana}}$

$303.75 \times 1 = 135x$ Find cross products.

$303.75 = 135x$

$\frac{303.75}{135} = x$

$2.25 = x$

Therefore, one bandana costs \$2.25.

▶ Movement

When working with movement problems, it is important to use the following formula:

(Rate)(Time) = Distance

Example

A boat traveling at 45 mph traveled around a lake in 0.75 hours less than a boat traveling at 30 mph. What was the distance around the lake?

First, write what is known and unknown.

Unknown = time for Boat 2, traveling 30 mph to go around the lake = x
Known = time for Boat 1, traveling 45 mph to go around the lake = $x - 0.75$
Then, use the formula (Rate)(Time) = Distance to write an equation. The distance around the lake does not change for either boat, so you can make the two expressions equal to each other:
(Boat 1 rate)(Boat 1 time) = Distance around lake
(Boat 2 rate)(Boat 2 time) = Distance around lake
Therefore:
(Boat 1 rate)(Boat 1 time) = (Boat 2 rate)(Boat 2 time)
$(45)(x - 0.75) = (30)(x)$
$45x - 33.75 = 30x$
$45x - 33.75 - 45x = 30x - 45x$
$-\frac{33.75}{15} = -\frac{15x}{15}$
$-2.25 = -x$
$2.25 = x$
Remember: x represents the time it takes Boat 2 to travel around the lake. We need to plug it into the formula to determine the distance around the lake:
(Rate)(Time) = Distance
(Boat 2 Rate)(Boat 2 Time) = Distance
(30)(2.25) = Distance
67.5 = Distance
The distance around the lake is 67.5 miles.

Practice Question

Priscilla rides her bike to school at an average speed of 8 miles per hour. She rides her bike home along the same route at an average speed of 4 miles per hour. Priscilla rides a total of 3.2 miles round-trip. How many hours does it take her to ride round-trip?

a. 0.2
b. 0.4
c. 0.6
d. 0.8
e. 2

Answer

c. Let's determine the time it takes Priscilla to complete each leg of the trip and then add the two times together to get the answer. Let's start with the trip from home to school:
Unknown = time to ride from home to school = x
Known = rate from home to school = 8 mph
Known = distance from home to school = total distance round-trip ÷ 2 = 3.2 miles ÷ 2 = 1.6 miles
Then, use the formula (Rate)(Time) = Distance to write an equation:
(Rate)(Time) = Distance
$8x = 1.6$

$\frac{8x}{8} = \frac{1.6}{8}$
$x = 0.2$
Therefore, Priscilla takes 0.2 hours to ride from home to school.
Now let's do the same calculations for her trip from school to home:
Unknown = time to ride from school to home = y
Known = rate from home to school = 4 mph
Known = distance from school to home = total distance round-trip ÷ 2 = 3.2 miles ÷ 2 = 1.6 miles
Then, use the formula (Rate)(Time) = Distance to write an equation:
(Rate)(Time) = Distance
$4x = 1.6$
$\frac{4x}{4} = \frac{1.6}{4}$
$x = 0.4$
Therefore, Priscilla takes 0.4 hours to ride from school to home.
Finally add the times for each leg to determine the total time it takes Priscilla to complete the round trip:
$0.4 + 0.2 = 0.6$ hours
It takes Priscilla 0.6 hours to complete the round-trip.

▶ Work-Output Problems

Work-output problems deal with the rate of work. In other words, they deal with how much work can be completed in a certain amount of time. The following formula can be used for these problems:

(rate of work)(time worked) = part of job completed

Example

Ben can build two sand castles in 50 minutes. Wylie can build two sand castles in 40 minutes. If Ben and Wylie work together, how many minutes will it take them to build one sand castle?
Since Ben can build two sand castles in 60 minutes, his rate of work is $\frac{\text{2 sand castles}}{\text{60 minutes}}$ or $\frac{\text{1 sand castle}}{\text{30 minutes}}$. Wylie's rate of work is $\frac{\text{2 sand castles}}{\text{40 minutes}}$ or $\frac{\text{1 sand castle}}{\text{20 minutes}}$.
To solve this problem, making a chart will help:

	RATE	TIME	=	PART OF JOB COMPLETED
Ben	$\frac{1}{30}$	x	=	1 sand castle
Wylie	$\frac{1}{20}$	x	=	1 sand castle

Since Ben and Wylie are both working together on one sand castle, you can set the equation equal to one:
(Ben's rate)(time) + (Wylie's rate)(time) = 1 sand castle
$\frac{1}{30}x + \frac{1}{20}x = 1$

Now solve by using 60 as the LCD for 30 and 20:

$\frac{1}{30}x + \frac{1}{20}x = 1$

$\frac{2}{60}x + \frac{3}{60}x = 1$

$\frac{5}{60}x = 1$

$\frac{5}{60}x \times 60 = 1 \times 60$

$5x = 60$

$x = 12$

Thus, it will take Ben and Wylie 12 minutes to build one sand castle.

Practice Question

Ms. Walpole can plant nine shrubs in 90 minutes. Mr. Saum can plant 12 shrubs in 144 minutes. If Ms. Walpole and Mr. Saum work together, how many minutes will it take them to plant two shrubs?

a. $\frac{60}{11}$

b. 10

c. $\frac{120}{11}$

d. 11

e. $\frac{240}{11}$

Answer

c. Ms. Walpole can plant 9 shrubs in 90 minutes, so her rate of work is $\frac{9 \text{ shrubs}}{90 \text{ minutes}}$ or $\frac{1 \text{ shrub}}{10 \text{ minutes}}$. Mr. Saum's rate of work is $\frac{12 \text{ shrubs}}{144 \text{ minutes}}$ or $\frac{1 \text{ shrub}}{12 \text{ minutes}}$.

To solve this problem, making a chart will help:

	RATE	TIME	=	PART OF JOB COMPLETED
Ms. Walpole	$\frac{1}{10}$	x	=	1 shrub
Mr. Saum	$\frac{1}{12}$	x	=	1 shrub

Because both Ms. Walpole and Mr. Saum are working together on two shrubs, you can set the equation equal to two:

(Ms. Walpole's rate)(time) + (Mr. Saum's rate)(time) = 2 shrubs

$\frac{1}{10}x + \frac{1}{12}$

$x = 2$

Now solve by using 60 as the LCD for 10 and 12:

$\frac{1}{10}x + \frac{1}{12}x = 2$

$\frac{6}{60}x + \frac{5}{60}x = 2$

$\frac{11}{60}x = 2$

$\frac{11}{60}x \times 60 = 2 \times 60$

$11x = 120$

$x = \frac{120}{11}$

Thus, it will take Ms. Walpole and Mr. Saum $\frac{120}{11}$ minutes to plant two shrubs.

▶ Special Symbols Problems

Some SAT questions invent an operation symbol that you won't recognize. Don't let these symbols confuse you. These questions simply require you to make a substitution based on information the question provides. Be sure to pay attention to the placement of the variables and operations being performed.

Example

Given $p \lozenge q = (p \times q + 4)^2$, find the value of $2 \lozenge 3$.

Fill in the formula with 2 replacing p and 3 replacing q.

$(p \times q + 4)^2$

$(2 \times 3 + 4)^2$

$(6 + 4)^2$

$(10)^2$

$= 100$

So, $2 \lozenge 3 = 100$.

Example

If 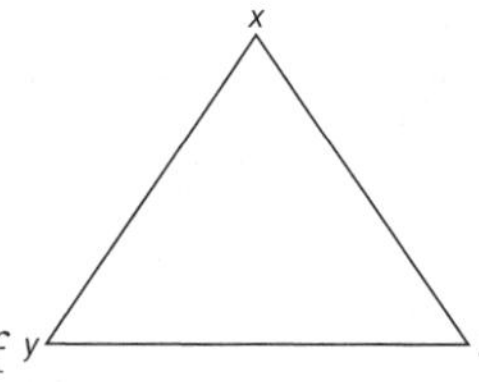

$= \frac{x+y+z}{x} + \frac{x+y+z}{y} + \frac{x+y+z}{z}$, then what is the value of 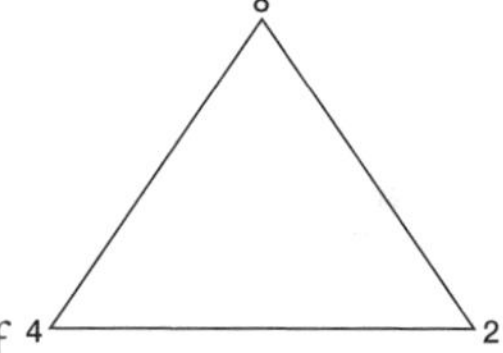

Fill in the variables according to the placement of the numbers in the triangular figure: $x = 8$, $y = 4$, and $z = 2$.

$\frac{8+4+2}{8} + \frac{8+4+2}{4} + \frac{8+4+2}{2}$

$\frac{14}{8} + \frac{14}{4} + \frac{14}{2}$ LCD is 8.

$\frac{14}{8} + \frac{28}{8} + \frac{56}{8}$ Add.

$\frac{98}{8}$ Simplify.

$\frac{49}{4}$

Answer: $\frac{49}{4}$

Practice Question

The operation $c \,\Omega\, d$ is defined by $c \,\Omega\, d = d^{c+d} \times d^{c-d}$. What value of d makes $2 \,\Omega\, d$ equal to 81?

a. 2
b. 3
c. 9
d. 20.25
e. 40.5

Answer

b. If $c \,\Omega\, d = d^{c+d} \times d^{c-d}$, then $2 \,\Omega\, d = d^{2+d} \times d^{2-d}$. Solve for d when $2 \,\Omega\, d = 81$:

$d^{2+d} \times d^{2-d} = 81$
$d^{(2+d)+(2-d)} = 81$
$d^{2+2+d-d} = 81$
$d^4 = 81$
$\sqrt{d^4} = \sqrt{81}$
$d^2 = 9$
$\sqrt{d^2} = \sqrt{9}$
$d = 3$

Therefore, $d = 3$ when $2 \,\Omega\, d = 81$.

▶ The Counting Principle

Some questions ask you to determine the number of outcomes possible in a given situation involving different choices.

For example, let's say a school is creating a new school logo. Students have to vote on one color for the background and one color for the school name. They have six colors to choose from for the background and eight colors to choose from for the school name. How many possible combinations of colors are possible?

The quickest method for finding the answer is to use **the counting principle.** Simply multiply the number of possibilities from the first category (six background colors) by the number of possibilities from the second category (eight school name colors):

$6 \times 8 = 48$

Therefore, there are 48 possible color combinations that students have to choose from.

Remember: When determining the number of outcomes possible when combining one out of x choices in one category and one out of y choices in a second category, simply multiply $x \times y$.

Practice Question

At an Italian restaurant, customers can choose from one of nine different types of pasta and one of five different types of sauce. How many possible combinations of pasta and sauce are possible?

a. $\frac{9}{5}$
b. 4
c. 14
d. 32
e. 45

Answer

e. You can use the counting principle to solve this problem. The question asks you to determine the number of combinations possible when combining one out of nine types of pasta and one out of five types of sauce. Therefore, multiply $9 \times 5 = 45$. There are 45 total combinations possible.

▶ Permutations

Some questions ask you to determine the number of ways to arrange n items in all possible groups of r items. For example, you may need to determine the total number of ways to arrange the letters *ABCD* in groups of two letters. This question involves four items to be arranged in groups of two items. Another way to say this is that the question is asking for the number of **permutations** it's possible make of a group with two items from a group of four items. Keep in mind when answering permutation questions that *the order of the items matters.* In other words, using the example, both *AB* and *BA* must be counted.

To solve permutation questions, you must use a special formula:

$${}_nP_r = \frac{n!}{(n-r)!}$$

P = number of permutations
n = the number of items
r = number of items in each permutation

Let's use the formula to answer the problem of arranging the letters *ABCD* in groups of two letters.

the number of items (n) = 4
number of items in each permutation (r) = 2

Plug in the values into the formula:

$${}_nP_r = \frac{n!}{(n-r)!}$$

$${}_4P_2 = \frac{4!}{(4-2)!}$$

$${}_4P_2 = \frac{4!}{2!}$$

$_4P_2 = \frac{4 \times 3 \times 2 \times 1}{2 \times 1}$ Cancel out the 2×1 from the numerator and denominator.

$_4P_2 = 4 \times 3$

$_4P_2 = 12$

Therefore, there are 12 ways to arrange the letters *ABCD* in groups of two:

AB	*AC*	*AD*	*BA*	*BC*	*BD*
CA	*CB*	*CD*	*DA*	*DB*	*DC*

Practice Question

Casey has four different tickets to give away to friends for a play she is acting in. There are eight friends who want to use the tickets. How many different ways can Casey distribute four tickets among her eight friends?

a. 24
b. 32
c. 336
d. 1,680
e. 40,320

Answer

d. To answer this permutation question, you must use the formula $_nP_r = \frac{n!}{(n-r)!}$, where n = the number of friends = 8 and r = the number of tickets that the friends can use = 4. Plug the numbers into the formula:

$_nP_r = \frac{n!}{(n-r)!}$

$_8P_4 = \frac{8!}{(8-4)!}$

$_8P_4 = \frac{8!}{4!}$

$_8P_4 = \frac{8 \times 7 \times 6 \times 5 \times 4 \times 3 \times 2 \times 1}{4 \times 3 \times 2 \times 1}$ Cancel out the $4 \times 3 \times 2 \times 1$ from the numerator and denominator.

$_8P_4 = 8 \times 7 \times 6 \times 5$

$_8P_4 = 1{,}680$

Therefore, there are 1,680 permutations of friends that she can give the four different tickets to.

▶ Combinations

Some questions ask you to determine the number of ways to arrange n items in groups of r items without repeated items. In other words, *the order of the items doesn't matter.* For example, to determine the number of ways to arrange the letters *ABCD* in groups of two letters in which the order doesn't matter, you would count only *AB*, not both *AB* and *BA*. These questions ask for the total number of **combinations** of items.

To solve combination questions, use this formula:

$${}_nC_r = \frac{{}_nP_r}{r!} = \frac{n!}{(n-r)!r!}$$

C = number of combinations

n = the number of items

r = number of items in each permutation

For example, to determine the number of three-letter combinations from a group of seven letters (*ABCDEFGH*), use the following values: $n = 7$ and $r = 3$.

Plug in the values into the formula:

$${}_7C_3 = \frac{n!}{(n-r)!r!} = \frac{7!}{(7-3)!3!} = \frac{7!}{4!3!} = \frac{7 \times 6 \times 5 \times 4 \times 3 \times 2 \times 1}{(4 \times 3 \times 2 \times 1)(3!)} = \frac{7 \times 6 \times 5}{3 \times 2 \times 1} = \frac{210}{6} = 35$$

Therefore there are 35 three-letter combinations from a group of seven letters.

Practice Question

A film club has five memberships available. There are 12 people who would like to join the club. How many combinations of the 12 people could fill the five memberships?

a. 60

b. 63

c. 792

d. 19,008

e. 95,040

Answer

c. The order of the people doesn't matter in this problem, so it is a combination question, not a permutation question. Therefore we can use the formula ${}_nC_r = \frac{n!}{(n-r)!r!}$, where n = the number of people who want the membership = 12 and r = the number of memberships = 5.

$${}_nC_r = \frac{n!}{(n-r)!r!}$$

$${}_{12}C_5 = \frac{12!}{(12-5)!5!}$$

$${}_{12}C_5 = \frac{12!}{7!5!}$$

$${}_{12}C_5 = \frac{12 \times 11 \times 10 \times 9 \times 8 \times 7 \times 6 \times 5 \times 4 \times 3 \times 2 \times 1}{(7 \times 6 \times 5 \times 4 \times 3 \times 2 \times 1)5!}$$

$${}_{12}C_5 = \frac{12 \times 11 \times 10 \times 9 \times 8}{5 \times 4 \times 3 \times 2 \times 1}$$

$${}_{12}C_5 = \frac{95{,}040}{120}$$

$${}_{12}C_5 = 792$$

Therefore, there are 792 different combinations of 12 people to fill five memberships.

► Probability

Probability measures the likelihood that a specific event will occur. Probabilities are expressed as fractions. To find the probability of a specific outcome, use this formula:

Probability of an event $= \frac{\text{number of specific outcomes}}{\text{total number of possible outcomes}}$

Example

If a hat contains nine white buttons, five green buttons, and three black buttons, what is the probability of selecting a green button without looking?

Probability $= \frac{\text{number of specific outcomes}}{\text{total number of possible outcomes}}$

Probability $= \frac{\text{number of green buttons}}{\text{total number of buttons}}$

Probability $= \frac{5}{9+5+3}$

Probability $= \frac{5}{17}$

Therefore, the probability of selecting a green button without looking is $\frac{5}{17}$.

Practice Question

A box of DVDs contains 13 comedies, four action movies, and 15 thrillers. If Brett selects a DVD from the box without looking, what is the probability he will pick a comedy?

a. $\frac{4}{32}$
b. $\frac{13}{32}$
c. $\frac{15}{32}$
d. $\frac{13}{15}$
e. $\frac{13}{4}$

Answer

b. Probability is $\frac{\text{number of specific outcomes}}{\text{total number of possible outcomes}}$. Therefore, you can set up the following fraction:

$\frac{\text{number of comedy DVDs}}{\text{total number of DVDs}} = \frac{13}{13+4+15} = \frac{13}{32}$

Therefore, the probability of selecting a comedy DVD is $\frac{13}{32}$.

Multiple Probabilities

To find the probability that one of two or more mutually exclusive events will occur, add the probabilities of each event occurring. For example, in the previous problem, if we wanted to find the probability of drawing either a green or black button, we would add the probabilities together.

The probability of drawing a green button $= \frac{5}{17}$.
The probability of drawing a black button $= \frac{\text{number of black buttons}}{\text{total number of buttons}} = \frac{3}{9+5+3} = \frac{3}{17}$.
So the probability for selecting either a green or black button $= \frac{5}{17} + \frac{3}{17} = \frac{8}{17}$.

Practice Question

At a farmers' market, there is a barrel filled with apples. In the barrel are 40 Fuji apples, 24 Gala apples, 12 Red Delicious apples, 24 Golden Delicious, and 20 McIntosh apples. If a customer reaches into the barrel and selects an apple without looking, what is the probability that she will pick a Fuji or a McIntosh apple?

a. $\frac{1}{6}$
b. $\frac{1}{3}$
c. $\frac{2}{5}$
d. $\frac{1}{2}$
e. $\frac{3}{5}$

Answer

d. This problem asks you to find the probability that two events will occur (picking a Fuji apple or picking a McIntosh apple), so you must add the probabilities of each event. So first find the probability that someone will pick a Fuji apple:

the probability of picking a Fuji apple =

$\frac{\text{number of Fuji apples}}{\text{total number of apples}} =$

$\frac{40}{40 + 24 + 12 + 24 + 20} =$

$\frac{40}{120}$

Now find the probability that someone will pick a McIntosh apple:
the probability of picking a McIntosh apple =

$\frac{\text{number of McIntosh apples}}{\text{total number of apples}} =$

$\frac{20}{40 + 24 + 12 + 24 + 20} =$

$\frac{20}{120}$

Now add the probabilities together:

$\frac{40}{120} + \frac{20}{120} = \frac{60}{120} = \frac{1}{2}$

The probability that someone will pick a Fuji apple or a McIntosh is $\frac{1}{2}$.

Helpful Hints about Probability

- If an event is certain to occur, its probability is 1.
- If an event is certain *not* to occur, its probability is 0.
- You can find the probability of an unknown event if you know the probability of all other events occurring. Simply add the known probabilities together and subtract the result from 1. For example, let's say there is a bag filled with red, orange, and yellow buttons. You want to know the probability that you will pick a red button from a bag, but you don't know how many red buttons there are. However, you do know that the probability of picking an orange button is $\frac{3}{20}$ and the probability of picking a yellow button is $\frac{14}{20}$. If you add these probabilities together, you know the probability that you will pick an orange or yellow button: $\frac{3}{20} + \frac{16}{20} = \frac{19}{20}$. This probability, $\frac{19}{20}$, is also the probability that you *won't* pick a *red* button. Therefore, if you subtract $1 - \frac{19}{20}$, you will know the probability that you *will* pick a red button. $1 - \frac{19}{20} = \frac{1}{20}$. Therefore, the probability of choosing a red button is $\frac{1}{20}$.

Practice Question

Angie ordered 75 pizzas for a party. Some are pepperoni, some are mushroom, some are onion, some are sausage, and some are olive. However, the pizzas arrived in unmarked boxes, so she doesn't know which box contains what kind of pizza. The probability that a box contains a pepperoni pizza is $\frac{1}{15}$, the probability that a box contains a mushroom pizza is $\frac{2}{15}$, the probability that a box contains an onion pizza is $\frac{16}{75}$, and the probability that a box contains a sausage pizza is $\frac{8}{25}$. If Angie opens a box at random, what is the probability that she will find an olive pizza?

a. $\frac{2}{15}$
b. $\frac{1}{5}$
c. $\frac{4}{15}$
d. $\frac{11}{15}$
e. $\frac{4}{5}$

Answer

c. The problem does not tell you the probability that a random box contains an olive pizza. However, the problem does tell you the probabilities of a box containing the other types of pizza. If you add together all those probabilities, you will know the probability that a box contains a pepperoni, a mushroom, an onion, or a sausage pizza. In other words, you will know the probability that a box does NOT contain an olive pizza:

pepperoni + mushroom + onion + sausage

$= \frac{1}{15} + \frac{2}{15} + \frac{16}{75} + \frac{8}{25}$ Use an LCD of 75.

$= \frac{5}{75} + \frac{10}{75} + \frac{16}{75} + \frac{24}{75}$

$= \frac{5}{75} + \frac{10}{75} + \frac{16}{75} + \frac{24}{75}$

$= \frac{55}{75}$

The probability that a box does NOT contain an olive pizza is $\frac{55}{75}$.

Now subtract this probability from 1:

$1 - \frac{55}{75} = \frac{75}{75} - \frac{55}{75} = \frac{20}{75} = \frac{4}{15}$

The probability of opening a box and finding an olive pizza is $\frac{4}{15}$.

CHAPTER

Practice Test 1

This practice test is a simulation of the three Math sections you will complete on the SAT. To receive the most benefit from this practice test, complete it as if it were the real SAT. So, take this practice test under test-like conditions: Isolate yourself somewhere you will not be disturbed; use a stopwatch; follow the directions; and give yourself only the amount of time allotted for each section.

When you are finished, review the answers and explanations that immediately follow the test. Make note of the kinds of errors you made and review the appropriate skills and concepts before taking another practice test.

▶ Section 1

1. ⓐ ⓑ ⓒ ⓓ ⓔ
2. ⓐ ⓑ ⓒ ⓓ ⓔ
3. ⓐ ⓑ ⓒ ⓓ ⓔ
4. ⓐ ⓑ ⓒ ⓓ ⓔ
5. ⓐ ⓑ ⓒ ⓓ ⓔ
6. ⓐ ⓑ ⓒ ⓓ ⓔ
7. ⓐ ⓑ ⓒ ⓓ ⓔ
8. ⓐ ⓑ ⓒ ⓓ ⓔ
9. ⓐ ⓑ ⓒ ⓓ ⓔ
10. ⓐ ⓑ ⓒ ⓓ ⓔ
11. ⓐ ⓑ ⓒ ⓓ ⓔ
12. ⓐ ⓑ ⓒ ⓓ ⓔ
13. ⓐ ⓑ ⓒ ⓓ ⓔ
14. ⓐ ⓑ ⓒ ⓓ ⓔ
15. ⓐ ⓑ ⓒ ⓓ ⓔ
16. ⓐ ⓑ ⓒ ⓓ ⓔ
17. ⓐ ⓑ ⓒ ⓓ ⓔ
18. ⓐ ⓑ ⓒ ⓓ ⓔ
19. ⓐ ⓑ ⓒ ⓓ ⓔ
20. ⓐ ⓑ ⓒ ⓓ ⓔ

▶ Section 2

1. ⓐ ⓑ ⓒ ⓓ ⓔ
2. ⓐ ⓑ ⓒ ⓓ ⓔ
3. ⓐ ⓑ ⓒ ⓓ ⓔ
4. ⓐ ⓑ ⓒ ⓓ ⓔ
5. ⓐ ⓑ ⓒ ⓓ ⓔ
6. ⓐ ⓑ ⓒ ⓓ ⓔ
7. ⓐ ⓑ ⓒ ⓓ ⓔ
8. ⓐ ⓑ ⓒ ⓓ ⓔ

9.

10.

11.

12.

13.

14.

15.

16.

17.

18.

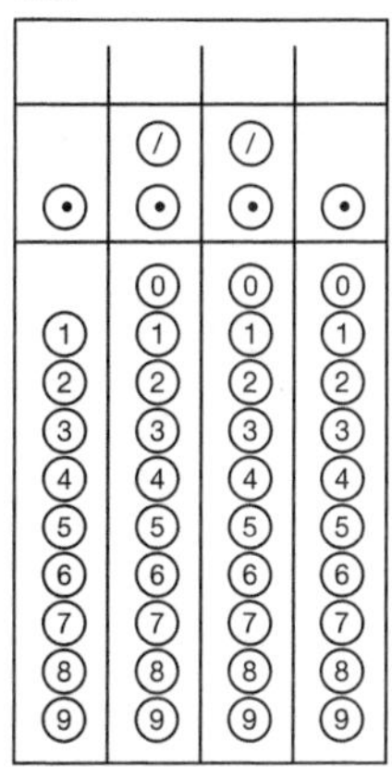

Section 3

1. (a) (b) (c) (d) (e)
2. (a) (b) (c) (d) (e)
3. (a) (b) (c) (d) (e)
4. (a) (b) (c) (d) (e)
5. (a) (b) (c) (d) (e)
6. (a) (b) (c) (d) (e)
7. (a) (b) (c) (d) (e)
8. (a) (b) (c) (d) (e)
9. (a) (b) (c) (d) (e)
10. (a) (b) (c) (d) (e)
11. (a) (b) (c) (d) (e)
12. (a) (b) (c) (d) (e)
13. (a) (b) (c) (d) (e)
14. (a) (b) (c) (d) (e)
15. (a) (b) (c) (d) (e)
16. (a) (b) (c) (d) (e)

▶ Section 1

1. If the expression $\frac{3}{2+x} = \frac{x-5}{2x}$, then one possible value of x could be

a. –1.
b. –2.
c. –5.
d. 1.
e. 2.

2.

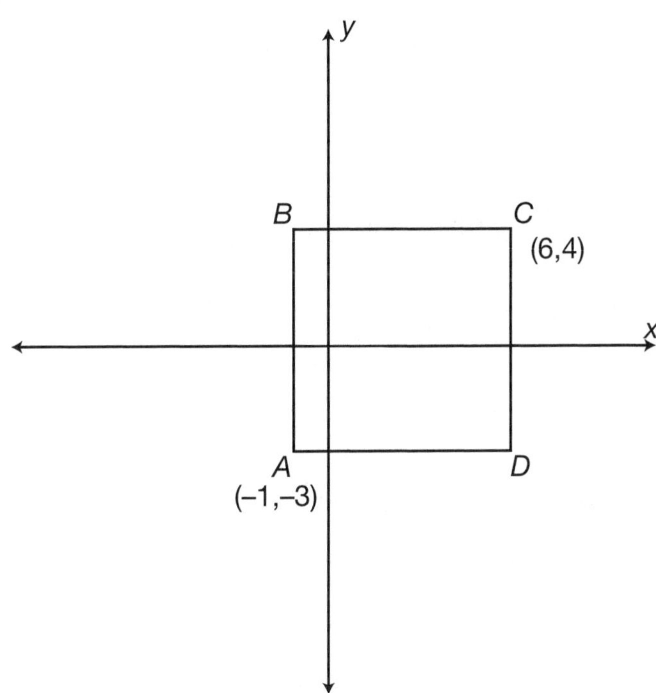

In the graph above, *ABCD* is a square. What are the coordinates of point *B*?

a. (–1,–4)
b. (–1,4)
c. (–1,6)
d. (–3,1)
e. (–3,4)

3. Line $y = \frac{2}{3}x - 5$ is perpendicular to line

a. $y = \frac{2}{3}x + 5$.
b. $y = 5 - \frac{2}{3}x$.
c. $y = -\frac{2}{3}x - 5$.
d. $y = \frac{3}{2}x - 5$.
e. $y = -\frac{3}{2}x + 5$.

4. If 30% of r is equal to 75% of s, what is 50% of s if $r = 30$?
   a. 4.5
   b. 6
   c. 9
   d. 12
   e. 15

5. A dormitory now houses 30 men and allows 42 square feet of space per man. If five more men are put into this dormitory, how much less space will each man have?
   a. 5 square feet
   b. 6 square feet
   c. 7 square feet
   d. 8 square feet
   e. 9 square feet

6. Rob has six songs on his portable music player. How many different four-song orderings can Rob create?
   a. 30
   b. 60
   c. 120
   d. 360
   e. 720

7. The statement "Raphael runs every Sunday" is always true. Which of the following statements is also true?
   a. If Raphael does not run, then it is not Sunday.
   b. If Raphael runs, then it is Sunday.
   c. If it is not Sunday, then Raphael does not run.
   d. If it is Sunday, then Raphael does not run.
   e. If it is Sunday, it is impossible to determine if Raphael runs.

8.

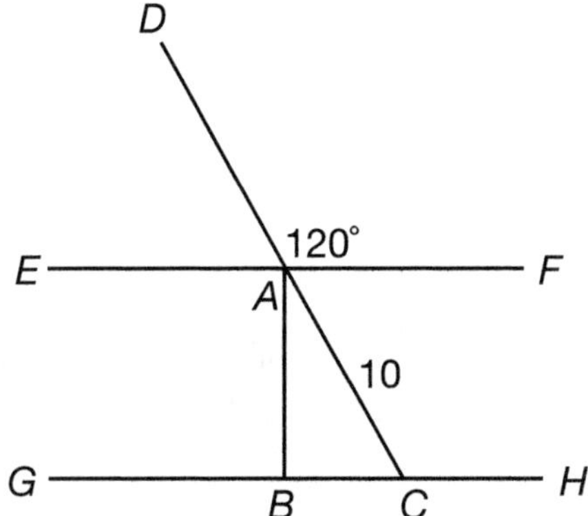

In the diagram above, lines *EF* and *GH* are parallel, and line *AB* is perpendicular to lines *EF* and *GH*. What is the length of line *AB*?

a. 5
b. $5\sqrt{2}$
c. $5\sqrt{3}$
d. $10\sqrt{2}$
e. $10\sqrt{3}$

9. The expression $\frac{(x^2 + 2x - 15)}{(x^2 + 4x - 21)}$ is equivalent to

a. $\frac{5}{7}$.
b. $x + 5$.
c. $\frac{x+5}{x+7}$.
d. $\frac{-5}{2x-7}$.
e. $\frac{2x-15}{4x-21}$.

10. The point (2,1) is the midpoint of a line with endpoints at (–5,3) and

a. (–3,4).
b. (–7,2).
c. (7,1).
d. (9,–1).
e. (–10,3).

11. Lindsay grows only roses and tulips in her garden. The ratio of roses to tulips in her garden is 5:6. If there are 242 total flowers in her garden, how many of them are tulips?

a. 22
b. 40
c. 110
d. 121
e. 132

12. It takes eight people 12 hours to clean an office. How long would it take six people to clean the office?

a. 9 hours
b. 15 hours
c. 16 hours
d. 18 hours
e. 24 hours

13. Greg has nine paintings. The Hickory Museum has enough space to display three of them. From how many different sets of three paintings does Greg have to choose?

a. 27
b. 56
c. 84
d. 168
e. 504

14. If the surface area of a cube is 384 cm^2, what is the volume of the cube?

a. 64 cm^3
b. 256 cm^3
c. 512 cm^3
d. 1,152 cm^3
e. 4,096 cm^3

15.

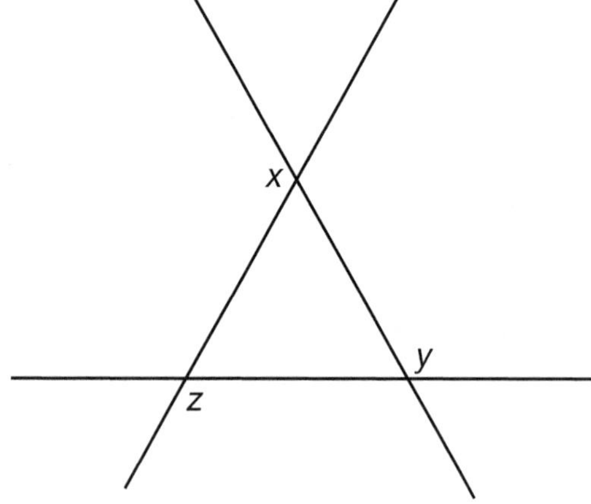

In the diagram above, what is the sum of the measures of the angles x, y, and z?

a. 180 degrees
b. 360 degrees
c. 540 degrees
d. 720 degrees
e. cannot be determined

16. Given the following figure with one tangent and one secant drawn to the circle, what is the measure of angle *ADB*?

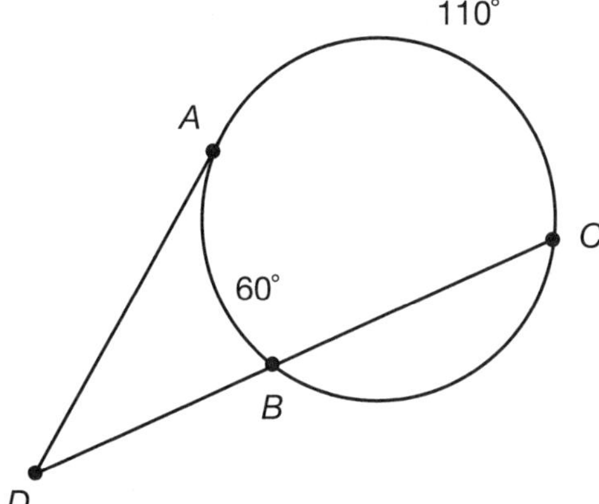

a. 50 degrees
b. 85 degrees
c. 60 degrees
d. 110 degrees
e. 25 degrees

17.

COST OF BALLONS	
QUANTITY	**PRICE PER BALLOON**
1	$1.00
10	$0.90
100	$0.75
1,000	$0.60

Balloons are sold according to the chart above. If a customer buys one balloon at a time, the cost is $1.00 per balloon. If a customer buys ten balloons at a time, the cost is $0.90 per balloon. If Carlos wants to buy 2,000 balloons, how much money does he save by buying 1,000 balloons at a time rather than ten balloons at a time?

a. $200
b. $300
c. $500
d. $600
e. $800

18. If $\frac{ab}{c} = d$, and a and c are doubled, what happens to the value of d?

a. The value of d remains the same.
b. The value of d is doubled.
c. The value of d is four times greater.
d. The value of d is halved.
e. The value of d is four times smaller.

19.

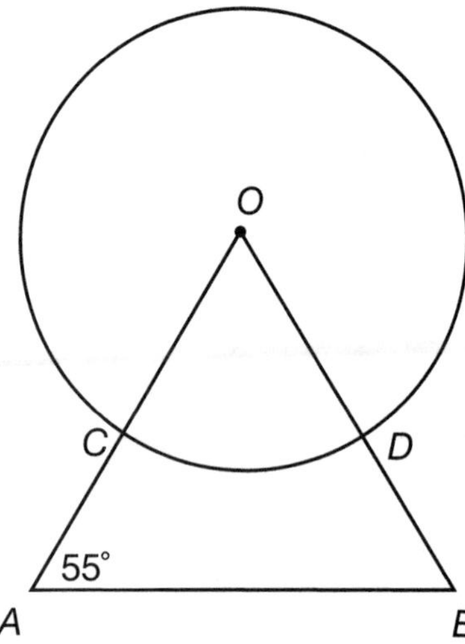

In the diagram above, line *OA* is congruent to line *OB*. What is the measure of arc *CD*?

a. 27.5 degrees
b. 55 degrees
c. 70 degrees
d. 110 degrees
e. 125 degrees

20. The expression $\frac{x\sqrt{32}}{\sqrt{4x}}$ is equivalent to

a. $2\sqrt{2}$.
b. $\frac{\sqrt{2}}{2}$.
c. $\frac{2\sqrt{2}}{\sqrt{x}}$.
d. $\frac{x\sqrt{2}}{\sqrt{x}}$.
e. $\frac{2x\sqrt{2}}{\sqrt{x}}$.

▶ Section 2

1. What is the next number in the series below?

3 16 6 12 12 8 ______

a. 4
b. 15
c. 20
d. 24
e. 32

2. The volume of a glass of water placed in the sun decreases by 20%. If there are 240 mL of water in the glass now, what was the original volume of water in the glass?

a. 192 mL
b. 260 mL
c. 288 mL
d. 300 mL
e. 360 mL

3. What is the tenth term of the pattern below?

$\frac{2}{3}, \frac{4}{9}, \frac{8}{27}, \frac{16}{81} \ldots$

a. $\frac{20}{30}$
b. $\frac{2^{10}}{3}$
c. $\frac{2}{3^{10}}$
d. $(\frac{2}{3})^{\frac{2}{3}}$
e. $(\frac{2}{3})^{10}$

4. How does the area of a rectangle change if both the base and the height of the original rectangle are tripled?

a. The area is tripled.
b. The area is six times larger.
c. The area is nine times larger.
d. The area remains the same.
e. The area cannot be determined.

5. The equation $y = \frac{x+6}{x^2+7x-18}$ is undefined when $x =$

a. –9.
b. –2.
c. –6.
d. 0.
e. 9.

6.

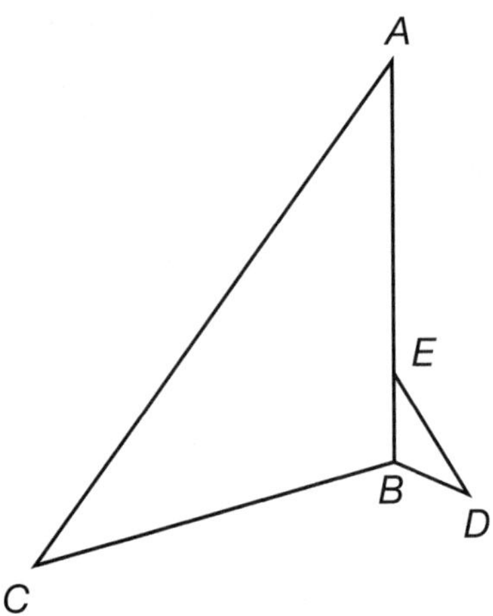

In the diagram above, angle A is congruent to angle BED, and angle C is congruent to angle D. If the ratio of the length of AB to the length of EB is 5:1, and the area of triangle $BED = 5a^2 + 10$, what is area of triangle ABC?

a. $5a^2 + 10$
b. $25a^2 + 50$
c. $25a^2 + 100$
d. $125a^2 + 250$
e. cannot be determined

7. The number p is greater than 0, a multiple of 6, and a factor of 180. How many possibilities are there for the value of p?

a. 7
b. 8
c. 9
d. 10
e. 11

8. If $g > 0$ and $h < 0$, which of the following is always positive?

a. gh
b. $g + h$
c. $g - h$
d. $|h| - |g|$
e. h^g

9. The length of a room is three more than twice the width of the room. The perimeter of the room is 66 feet. What is the length of the room?

10.

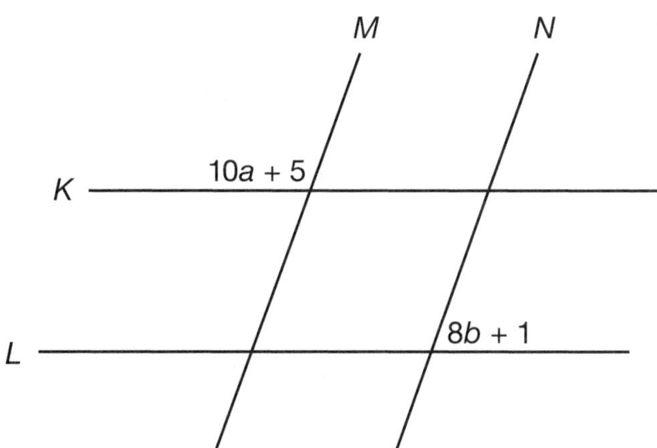

In the diagram above, lines K and L are parallel, and lines M and N are parallel. If $b = 8$, then $a = ?$

11. If $6x + 9y - 15 = -6$, what is the value of $-2x - 3y + 5$?

12. Find the measure of angle Z.

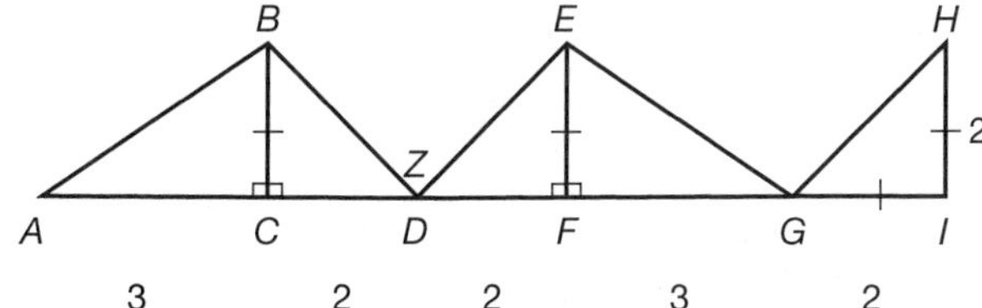

13. If the distance from point $(-2,m)$ to point $(4,-1)$ is 10 units, what is the positive value of m?

14. If $z^{\frac{2}{a}} = 9$, then $a = 3$ when $z = ?$

15. The length of a rectangular prism is four times the height of the prism and one-third the width of the prism. If the volume of the prism is 384 in^3, what is the width of the prism?

16. If $2a^2 + b = 10$ and $-\frac{b}{4} + 3a = 11$, what is the positive value of a?

17. Stephanie buys almonds at the grocery store for \$1.00 per pound. If she buys 4 pounds of almonds and pays a 5% tax on her purchase, what is Stephanie's total bill?

18. The ratio of the number of linear units in the circumference of a circle to the number of square units in the area of that circle is 2:5. What is the radius of the circle?

▶ Section 3

1. Which of the following number pairs is in the ratio 4:5?

a. $\frac{1}{4}, \frac{1}{5}$
b. $\frac{1}{5}, \frac{1}{4}$
c. $\frac{1}{5}, \frac{4}{5}$
d. $\frac{4}{5}, \frac{5}{4}$
e. $1, \frac{4}{5}$

2. When $x = -3$, the expression $-2x^2 + 3x - 7 =$

a. –34.
b. –27.
c. –16.
d. –10.
e. 2.

3. What is the slope of the line $-3y = 12x - 3$?

a. –4
b. –3
c. 1
d. 4
e. 12

4.

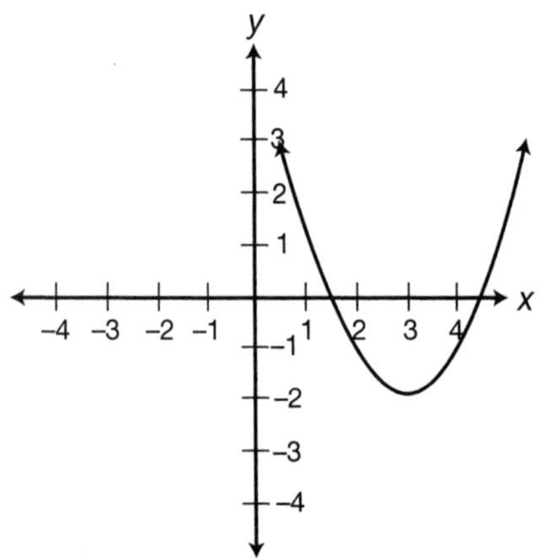

Which of the following could be the equation of the parabola shown above?

a. $y = (x + 3)^2 + 2$
b. $y = (x + 3)^2 - 2$
c. $y = (x - 3)^2 + 2$
d. $y = (x - 3)^2 - 2$
e. $y = (3x + 3)^2 - 2$

5. If $0.34 < x < 0.40$ and $\frac{5}{16} < x < \frac{9}{20}$, which of the following could be x?

a. $\frac{1}{3}$
b. $\frac{2}{5}$
c. $\frac{3}{8}$
d. $\frac{3}{7}$
e. $\frac{4}{9}$

6. A store prices a coat at $85. During a sale, the coat is sold at 20% off. After the sale, the store raises the price of the coat 10% over its sale price. What is the price of the coat now?

a. $18.70
b. $61.20
c. $68.00
d. $74.80
e. $93.50

7. The expression $4x^2 - 2x + 3$ is equal to 3 when $x = 0$ and when $x =$

a. $-\frac{1}{2}$.
b. $-\frac{1}{4}$.
c. $\frac{1}{8}$.
d. $\frac{1}{4}$.
e. $\frac{1}{2}$.

8. A spinner is divided into eight equal regions, labeled one through eight. If Jenna spins the wheel, what is the probability that she will spin a number that is less than four and greater than two?

a. $\frac{1}{8}$
b. $\frac{9}{32}$
c. $\frac{3}{8}$
d. $\frac{1}{2}$
e. $\frac{3}{4}$

9. The length of an edge of a cube is equal to half the height of a cylinder that has a volume of 160π cubic units. If the radius of the cylinder is 4 units, what is the surface area of the cube?

a. 64 square units
b. 96 square units
c. 100 square units
d. 125 square units
e. 150 square units

10. The function $m\#n$ is equal to $m^2 - n$. Which of the following is equivalent to $m\#(n\#m)$?

a. $-n$

b. $n^2 - m$

c. $m^2 + m - n^2$

d. $(m^2 - n)^2 - n$

e. $(n^2 - m)^2 - m$

11. Which of the following has the greatest value when $x = -\frac{1}{4}$?

a. x^{-1}

b. $-\frac{3}{8x}$

c. $4x + 3$

d. 16^x

e. $\frac{1}{81^x}$

12.

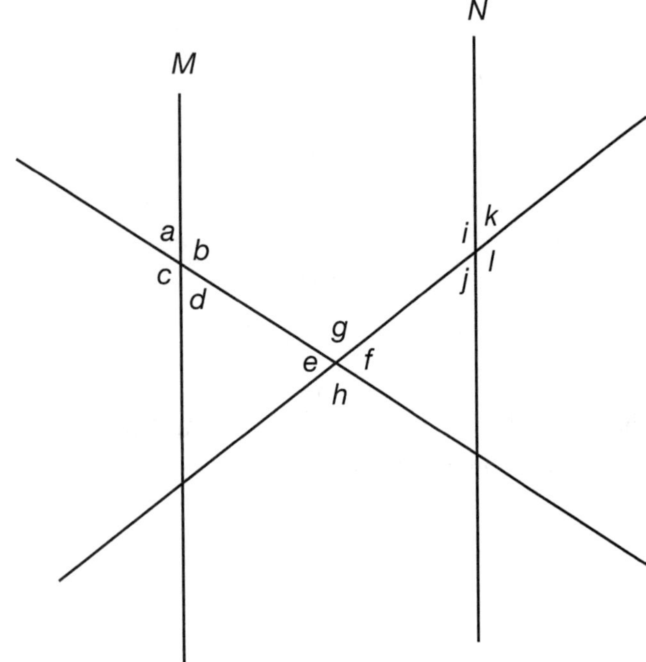

In the diagram above, lines M and N are parallel. All of the following are true EXCEPT

a. $a + b = j + l$.

b. $g = h$.

c. $c + f = f + b$.

d. $g + e + f + h = 360$.

e. $d + e = f + j$.

13. Melissa runs the 50-yard dash five times, with times of 5.4 seconds, 5.6 seconds, 5.4 seconds, 6.3 seconds, and 5.3 seconds. If she runs a sixth dash, which of the following would change the mean and mode of her scores, but not the median?

a. 5.3 seconds
b. 5.4 seconds
c. 5.5 seconds
d. 5.6 seconds
e. 6.3 seconds

14. If $x \neq 0$ and $y \neq 0$, $\dfrac{\frac{xy}{y} + xy}{\frac{xy}{x}} =$

a. $\frac{x}{y} + 1.$
b. $\frac{x}{y} + x.$
c. $\frac{x}{y} + y.$
d. $2xy.$
e. $y^2 + x.$

15.

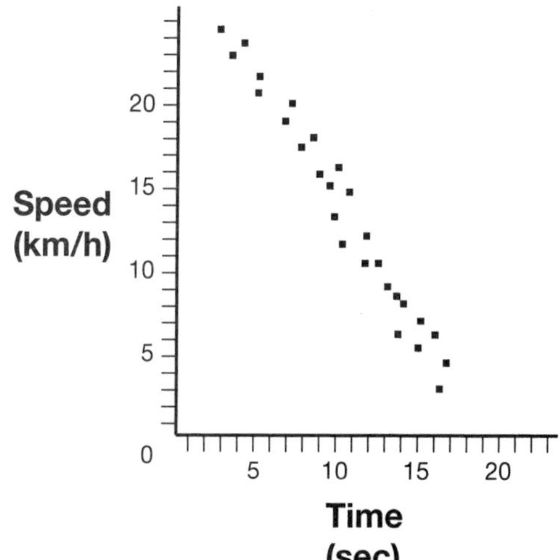

The scatterplot above shows the speeds of different runners over time. Which of the following could be the equation of the line of best fit?

a. $s = -2(t - 15)$
b. $s = -t + 25$
c. $s = -\frac{1}{2}(t - 10)$
d. $s = \frac{1}{2}(t + 20)$
e. $s = 2(t + 15)$

16.

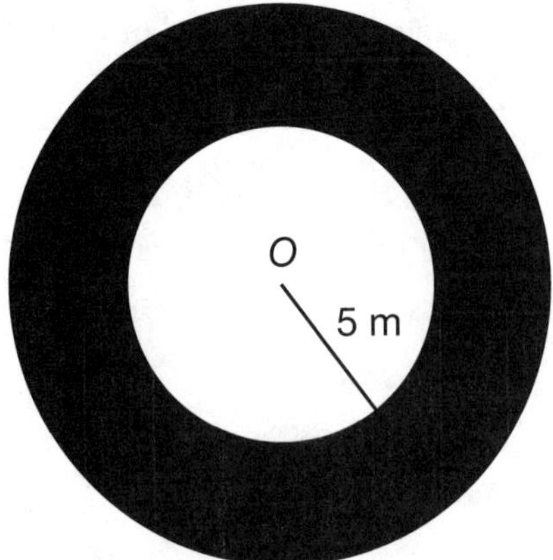

The radius of the outer circle shown above is 1.2 times greater than the radius of the inner circle. What is the area of the shaded region?

a. 6π m^2
b. 9π m^2
c. 25π m^2
d. 30π m^2
e. 36π m^2

▶ Answer Key

Section 1 Answers

1. a. Cross multiply and solve for x:

$3(2x) = (2 + x)(x - 5)$

$6x = x^2 - 3x - 10$

$x^2 - 9x - 10 = 0$

$(x - 10)(x + 1) = 0$

$x = 10, x = -1$

2. b. Point B is the same distance from the y-axis as point A, so the x-coordinate of point B is the same as the x-coordinate of point A: -1. Point B is the same distance from the x-axis as point C, so the y-coordinate of point B is the same as the y-coordinate of point C: 4. The coordinates of point B are $(-1,4)$.

3. e. Perpendicular lines have slopes that are negative reciprocals of each other. The slope of the line given is $\frac{2}{3}$. The negative reciprocal of $\frac{2}{3}$ is $-\frac{3}{2}$. Every line with a slope of $-\frac{3}{2}$ is perpendicular to the given line; $y = -\frac{3}{2}x + 5$ is perpendicular to $y = \frac{2}{3}x - 5$.

4. b. If $r = 30$, 30% of $r = (0.30)(3) = 9$. 9 is equal to 75% of s. If $0.75s = 9$, then $s = 12$. 50% of $s = (0.50)(12) = 6$.

5. b. 30 men × 42 square feet = 1,260 square feet of space; 1,260 square feet ÷ 35 men = 36 square feet; 42 – 36 = 6, so each man will have 6 less square feet of space.

6. d. The order of the four songs is important. The orderings A, B, C, D and A, C, B, D contain the same four songs, but in different orders. Both orderings must be counted. The number of six-choose-four orderings is equal to $(6)(5)(4)(3) = 360$.

7. a. The statement "Raphael runs every Sunday" is equivalent to "If it is Sunday, Raphael runs." The contrapositive of a true statement is also true. The contrapositive of "If it is Sunday, Raphael runs" is "If Raphael does not run, it is not Sunday."

8. c. Line AB is perpendicular to line BC, which makes triangle ABC a right triangle. Angles DAF and DCH are alternating angles—angles made by a pair of parallel lines cut by a transversal. Angle $DAF \cong$ angle DCH, therefore, angle DCH = 120 degrees. Angles DCH and ACB form a line. There are 180 degrees in a line, so the measure of angle $ACB = 180 - 120 = 60$ degrees. Triangle ABC is a 30-60-90 right triangle, which means that the length of the hypotenuse, AC, is equal to twice the length of the leg opposite the 30-degree angle, BC. Therefore, the length of BC is $\frac{10}{2}$, or 5. The length of the leg opposite the 60-degree angle, AB, is $\sqrt{3}$ times the length of the other leg, BC. Therefore, the length of AB is $5\sqrt{3}$.

9. c. Factor the numerator and denominator and cancel like factors:

$x^2 + 2x - 15 = (x + 5)(x - 3)$

$x^2 + 4x - 21 = (x + 7)(x - 3)$

Cancel the $(x - 3)$ term from the numerator and the denominator. The fraction reduces to $\frac{x+5}{x+7}$.

10. d. The midpoint of a line is equal to the average x-coordinates and the average y-coordinates of the line's endpoints:

$\frac{-5+x}{2} = 2, -5 + x = 4, x = 9$

$\frac{3+y}{2} = 1, 3 + y = 2, y = -1$

The other endpoint of this line is at $(9,-1)$.

11. e. The number of roses, $5x$, plus the number of tulips, $6x$, is equal to 242 total flowers: $5x + 6x = 242$, $11x = 242$, $x = 22$. There are $5(22) = 110$ roses and $6(22) = 132$ tulips in Lindsay's garden.

12. c. There is an inverse relationship between the number of people and the time needed to clean the office. Multiply the number of people by the hours needed to clean the office: $(8)(12) = 96$. Divide the total number of hours by the new number of people, 6: $\frac{96}{6} = 16$. It takes six people 16 hours to clean the office.

13. c. Be careful not to count the same set of three paintings more than once—order is not important. A nine-choose-three combination is equal to $\frac{(9)(8)(7)}{(3)(2)(1)} = \frac{504}{6} = 84$.

14. c. The surface area of a cube is equal to $6e^2$, where e is the length of one edge of the cube; $6e^2 = 384$ cm, $e^2 = 64$, $e = 8$ cm. The volume of a cube is equal to e^3; $(8 \text{ cm})^3 = 512 \text{ cm}^3$.

15. b. There are 180 degrees in a line: (x + (supplement of angle x)) + (y + (supplement of angle y)) + (z + (supplement of angle z)) = 540. The supplement of angle x, the supplement of angle y, and the supplement of angle z are the interior angles of a triangle. There are 180 degrees in a triangle, so those supplements sum to 180. Therefore, $x + y + z + 180 = 540$, and $x + y + z = 360$.

16. e. The measure of an angle in the exterior of a circle formed by a tangent and a secant is equal to half the difference of the intercepted arcs. The two intercepted arcs are $\overparen{AB}$, which is 60°, and $\overparen{AC}$, which is 110°. Find half of the difference of the two arcs; $\frac{1}{2}(110 - 60) = \frac{1}{2}(50) = 25°$.

17. d. If Carlos buys ten balloons, he will pay (10)($0.90) = $9. In order to total 2,000 balloons, Carlos will have to make this purchase $\frac{2{,}000}{10} = 200$ times. It will cost him a total of (200)($9) = $1,800. If Carlos buys 1,000 balloons, he will pay (1,000)($0.60) = $600. In order to total 2,000 balloons, Carlos will have to make this purchase $\frac{2{,}000}{1{,}000} = 2$ times. It will cost him a total of (2)($600) = $1,200. It will save Carlos $1,800 – $1,200 = $600 to buy the balloons 1,000 at a time.

18. a. If a and c are doubled, the fraction on the left side of the equation becomes $\frac{2ab}{2c}$. The fraction has been multiplied by $\frac{2}{2}$, which is equal to 1. Multiplying a fraction by 1 does not change its value; $\frac{2ab}{2c} = \frac{ab}{c} = d$. The value of d remains the same.

19. c. Triangle AOB is isosceles because line OA is congruent to line OB. Angles A and B are both 55 degrees, which means that angle $O = 180 - (55 + 55) = 70$ degrees. Angle O is a central angle and arc CD is its intercepted arc. A central angle and its intercepted arc are equal in measure, so the measure of arc CD is 70 degrees.

20. e. Simplify the numerator: $x\sqrt{32} = x\sqrt{16}\sqrt{2} = 4x\sqrt{2}$. Simplify the denominator: $\sqrt{4x} = \sqrt{4}\sqrt{x} = 2\sqrt{x}$. Divide the numerator and denominator by 2: $\frac{4x\sqrt{2}}{2\sqrt{x}} = \frac{2x\sqrt{2}}{\sqrt{x}}$.

Section 2 Answers

1. d. This series actually has two alternating sets of numbers. The first number is doubled, giving the third number. The second number has 4 subtracted from it, giving it the fourth number. Therefore, the blank space will be 12 doubled, or 24.

2. d. The original volume of water, x, minus 20% of x, $0.20x$, is equal to the current volume of water, 240 mL:

$x - 0.20x = 240$ mL

$0.8x = 240$ mL

$x = 300$ mL

3. e. Each term in the pattern is equal to the fraction $\frac{2}{3}$ raised to an exponent that is equal to the position of the term in the sequence. The first term in the sequence is equal to $(\frac{2}{3})^1$, the second term is equal to $(\frac{2}{3})^2$, and so on. Therefore, the tenth term in the sequence will be equal to $(\frac{2}{3})^{10}$.

4. c. Since both dimensions are tripled, there are two additional factors of 3. Therefore, the new area is $3 \times 3 = 9$ times as large as the original. For example, use a rectangle with a base of 5 and height of 6. The area is $5 \times 6 = 30$ square units. If you multiply the each side length by 3, the new dimensions are 15 and 18. The new area is 15×18, which is 270 square units. By comparing the new area with the original area, 270 square units is nine times larger than 30 square units; $30 \times 9 = 270$.

5. a. An equation is undefined when the value of a denominator in the equation is equal to zero. Set $x^2 + 7x - 18$ equal to zero and factor the quadratic to find its roots:

$x^2 + 7x - 18 = 0$

$(x + 9)(x - 2) = 0$

$x = -9, x = 2$

6. d. Triangles *ABC* and *BED* have two pairs of congruent angles. Therefore, the third pair of angles must be congruent, which makes these triangles similar. If the area of the smaller triangle, *BED*, is equal to $\frac{bh}{2}$, then the area of the larger triangle, *ABC*, is equal to $\frac{(5b)(5h)}{2}$ or $25(\frac{bh}{2})$. The area of triangle *ABC* is 25 times larger than the area of triangle *BED*. Multiply the area of triangle *BED* by 25: $25(5a^2 + 10) = 125a^2 + 250$.

7. b. The positive factors of 180 (the positive numbers that divide evenly into 180) are 1, 2, 3, 4, 5, 6, 9, 10, 12, 15, 18, 20, 30, 36, 45, 60, 90, and 180. Of these numbers, 8 (6, 12, 18, 30, 36, 60, 90, and 180) are multiples of 6.

8. c. A positive number minus a negative number will not only always be a positive number, but will also be a positive number greater than the first operand. gh will always be negative when one multiplicand is positive and the other is negative. $g + h$ will be positive when the absolute value of g is greater than the absolute value of h, but $g + h$ will be negative when the absolute value of g is less than the absolute value of h. $|h| - |g|$ will be positive when $|h|$ is greater than g, but $|h| - |g|$ will be negative when $|h|$ is less than g. h^g will be positive when g is an even, whole number, but negative when g is an odd, whole number.

9. 23 If x is the width of the room, then $3 + 2x$ is the length of the room. The perimeter is equal to $x + x + (3 + 2x) + (3 + 2x) = 66$; $6x + 6 = 66$; $6x = 60$; $x = 10$. The length of the room is equal to $2x + 3$, $2(10) + 3 = 23$ feet.

10. 11 The labeled angle formed by lines *M* and *K* and the supplement of the labeled angle formed by lines *L* and *N* are alternating angles. Therefore, they are congruent. The angle labeled $(10a + 5)$ and its supplement, which is equal to $(8b + 1)$, total 180 degrees: $(10a + 5) + (8b + 1) = 180$. If $b = 8$, then:

$(10a + 5) + (8(8) + 1) = 180$

$10a + 70 = 180$

$10a = 110$

$a = 11$

11. 2 The first expression, $6x + 9y - 15$, is -3 times the second expression, $-2x - 3y + 5$ (multiply each term in the second expression by -3 and you'd get the first expression). Therefore, the value of the first expression, -6, is -3 times the value of the second expression. So, you can find the value of the second expression by dividing the value of the first expression by -3: $\frac{-6}{-3} = 2$. The value of $-2x - 3y + 5$ (2) is just $\frac{-1}{3}$ times the value of $6x + 9y - 15$ (-6) since $-2x - 3y + 5$ itself is $-\frac{1}{3}$ times $6x + 9y - 15$.

12. 90 Triangle *DBC* and triangle *DEF* are isosceles right triangles, which means the measures of $\angle BDC$ and $\angle EDF$ both equal 45°; $180 - (m\angle BDC + m\angle EDF) = m\angle Z$; $180 - 90 = m\angle Z$; $m\angle Z = 90°$.

13. 7 First, use the distance formula to form an equation that can be solved for m:

Distance $= \sqrt{(x_2 - x_1)^2 + (y_2 - y_1)^2}$

$10 = \sqrt{(4 - (-2))^2 + ((-1) - m)^2}$

$10 = \sqrt{(6)^2 + (-1 - m)^2}$

$10 = \sqrt{36 + m^2 + 2m + 1}$

$10 = \sqrt{m^2 + 2m + 37}$

$100 = m^2 + 2m + 37$

$m^2 + 2m - 63 = 0$

Now, factor $m^2 + 2m - 63$:

$(m + 9)(m - 7) = 0$

$m = 7$, $m = -9$. The positive value of m is 7.

14. 27 Substitute 3 for a: $z^{\frac{2}{3}} = 9$. To solve for z, raise both sides of the equation to the power $\frac{3}{2}$: $z^{\frac{2}{3}\frac{3}{2}} = 9^{\frac{3}{2}}$, $z = \sqrt{9^3} = 3^3 = 27$.

15. 24 If the height of the prism is h, then the length of the prism is four times that, $4h$. The length is one-third of the width, so the width is three times the length: $12h$. The volume of the prism is equal to its length multiplied by its width multiplied by its height:

$(h)(4h)(12h) = 384$

$48h^3 = 384$

$h^3 = 8$

$h = 2$

The height of the prism is 2 in, the length of the prism is (2 in)(4) = 8 in, and the width of the prism is (8 in)(3) = 24 in.

16. 3 Solve $2a^2 + b = 10$ for b: $b = 10 - 2a^2$. Substitute $(10 - 2a^2)$ for b in the second equation and solve for a:

$-\frac{10-2a^2}{4} + 3a = 11$

$-10 + 2a^2 + 12a = 44$

$2a^2 + 12a - 54 = 0$

$(2a - 6)(a + 9) = 0$

$2a - 6 = 0, a = 3$

$a + 9 = 0, a = -9$

The positive value of a is 3.

17. 4.20 If one pound of almonds costs \$1.00, then 4 pounds of almonds costs 4(\$1.00) = \$4.00. If Stephanie pays a 5% tax, then she pays (\$4.00)(0.05) = \$0.20 in tax. Her total bill is \$4.00 + \$0.20 = \$4.20.

18. 5 The circumference of a circle = $2\pi r$ and the area of a circle = πr^2. If the ratio of the number of linear units in the circumference to the number of square units in the area is 2:5, then five times the circumference is equal to twice the area:

$5(2\pi r) = 2(\pi r^2)$

$10\pi r = 2\pi r^2$

$10r = 2r^2$

$5r = r^2$

$r = 5$

The radius of the circle is equal to 5.

Section 3 Answers

1. b. Two numbers are in the ratio 4:5 if the second number is $\frac{5}{4}$ times the value of the first number; $\frac{1}{4}$ is $\frac{5}{4}$ times the value of $\frac{1}{5}$.

2. a. Substitute –3 for x:

$-2(-3)^2 + 3(-3) - 7 = -2(9) - 9 - 7 = -18 - 16$

$= -34$

3. a. First, convert the equation to slope-intercept form: $y = mx + b$. Divide both sides of the equation by –3:

$\frac{-3y}{-3} = \frac{12x-3}{-3}$

$y = -4x + 1$

The slope of a line written in this form is equal to the coefficient of the x term. The coefficient of the x term is –4, so the slope of the line is –4.

4. d. The equation of a parabola with its turning point c units to the right of the y-axis is written as $y = (x - c)^2$. The equation of a parabola with its turning point d units below the x-axis is written as $y = x^2 - d$. The parabola shown has its turning point three units to the right of the y-axis and two units below the x-axis, so its equation is $y = (x - 3)^2 - 2$. Alternatively, you can plug the coordinates of the vertex of the parabola, (3,–2), into each equation. The only equation that holds true is choice **d:** $y = (x - 3)^2 - 2$, $-2 = (3 - 3)^2 - 2$, $-2 = 0^2 - 2$, $-2 = -2$.

5. c. $\frac{5}{16} = 0.3125$ and $\frac{9}{20} = 0.45$; $\frac{3}{8} = 0.375$, which is between 0.34 and 0.40, and between 0.3125 and 0.45.

6. d. 20% of \$85 = (0.20)(\$85) = \$17. While on sale, the coat is sold for \$85 – \$17 = \$68; 10% of \$68 = (0.10)(\$68) = \$6.80. After the sale, the coat is sold for \$68 + \$6.80 = \$74.80.

7. e. Set the expression $4x^2 - 2x + 3$ equal to 3 and solve for x:

$4x^2 - 2x + 3 = 3$

$4x^2 - 2x + 3 - 3 = 3 - 3$

$4x^2 - 2x = 0$

$4x(x - \frac{1}{2}) = 0$

$x = 0, x = \frac{1}{2}$

8. a. There are three numbers on the wheel that are less than four (1, 2, 3), but only one of those numbers (3) is greater than two. The probability of Jenna spinning a number that is both less than 4 and greater than 2 is $\frac{1}{8}$.

9. e. The volume of a cylinder is equal to πr^2h. The volume of the cylinder is 160π and its radius is 4. Therefore, the height of the cylinder is equal to:

$160\pi = \pi(4)^2h$

$160 = 16h$

$h = 10$

The length of an edge of the cube is equal to half the height of the cylinder. The edge of the cube is 5 units. The surface area of a cube is equal to $6e^2$, where e is the length of an edge of the cube. The surface area of the cube $= 6(5)^2 = 6(25) =$ 150 square units.

10. c. $m\#n$ is a function definition. The problem is saying "$m\#n$" is the same as "$m^2 - n$". If $m\#n$ is $m^2 - n$, then $n\#m$ is $n^2 - m$. So, to find $m\#(n\#m)$, replace $(n\#m)$ with the value of $(n\#m)$, which is $n^2 - m$: $m\#(n^2 - m)$.

Now, use the function definition again. The function definition says "take the value before the # symbol, square it, and subtract the value after the # symbol": m squared is m^2, minus the second term, $(n^2 - m)$, is equal to $m^2 - (n^2 - m) = m^2 - n^2 + m$.

11. e. $x^{-1} = \frac{1}{x} = \frac{1}{-\frac{1}{4}} = -4; -\frac{3}{8x} = -\frac{3}{8(-\frac{1}{4})} = \frac{3}{2}$. $4x + 3 = 4(-\frac{1}{4}) + 3 = -1 + 3 = 2$; $16^x = 16^{-\frac{1}{4}} = \frac{1}{16^{\frac{1}{4}}} = \frac{1}{2}$; $\frac{1}{81^x} = \frac{1}{81^{-\frac{1}{4}}} = 81^{\frac{1}{4}} = 3$.

12. e. Angles e and f are vertical angles, so angle $e \cong$ angle f. However, angle d and angle j are not alternating angles. These angles are formed by different transversals. It cannot be stated that angle $d \cong$ angle j, therefore, it cannot be stated that $d + e = f + j$.

13. a. Melissa's mean time for the first five dashes is $\frac{5.4 + 5.6 + 5.4 + 6.3 + 5.3}{5} = \frac{28}{5} = 5.6$. Her times, in order from least to greatest, are: 5.3, 5.4, 5.4, 5.6, and 6.3. The middle score, or median, is 5.4. The number that appears most often, the mode, is 5.4. A score of 5.3 means that the mean will decrease and that the mode will no longer be 5.4 alone. The mode will now be 5.3 and 5.4. The median, however, will remain 5.4.

14. b. $\frac{\frac{xy}{y} + xy}{\frac{xy}{x}} = (\frac{xy}{y} + xy)(\frac{x}{xy}) = \frac{x}{y} + x$

15. a. If a straight line were drawn through as many of the plotted points as possible, it would have a negative slope. The line slopes more sharply than the line $y = -x$ (a line with a slope of -1), so the line would have a slope more negative than -1. The line would also have a y-intercept well above the x-axis. The only equation given with a slope more negative than -1 is $s = -2(t - 15)$.

16. b. The area of a circle is equal to πr^2. The radius of the inner circle is 5 m; therefore, the area of the inner circle is 25π m^2. The radius of the outer circle is $(1.2)(5) = 6$ m; therefore, the area of the outer circle is 36π. Subtract the area of the inner circle from the area of the outer circle: $36\pi - 25\pi = 9\pi$ m^2.

CHAPTER

10 Practice Test 2

This practice test is a simulation of the three Math sections you will complete on the SAT. To receive the most benefit from this practice test, complete it as if it were the real SAT. So take this practice test under test-like conditions: Isolate yourself somewhere you will not be disturbed; use a stopwatch; follow the directions; and give yourself only the amount of time allotted for each section.

When you are finished, review the answers and explanations that immediately follow the test. Make note of the kinds of errors you made and review the appropriate skills and concepts before taking another practice test.

► Section 1

1. ⓐ ⓑ ⓒ ⓓ ⓔ
2. ⓐ ⓑ ⓒ ⓓ ⓔ
3. ⓐ ⓑ ⓒ ⓓ ⓔ
4. ⓐ ⓑ ⓒ ⓓ ⓔ
5. ⓐ ⓑ ⓒ ⓓ ⓔ
6. ⓐ ⓑ ⓒ ⓓ ⓔ
7. ⓐ ⓑ ⓒ ⓓ ⓔ
8. ⓐ ⓑ ⓒ ⓓ ⓔ
9. ⓐ ⓑ ⓒ ⓓ ⓔ
10. ⓐ ⓑ ⓒ ⓓ ⓔ
11. ⓐ ⓑ ⓒ ⓓ ⓔ
12. ⓐ ⓑ ⓒ ⓓ ⓔ
13. ⓐ ⓑ ⓒ ⓓ ⓔ
14. ⓐ ⓑ ⓒ ⓓ ⓔ
15. ⓐ ⓑ ⓒ ⓓ ⓔ
16. ⓐ ⓑ ⓒ ⓓ ⓔ
17. ⓐ ⓑ ⓒ ⓓ ⓔ
18. ⓐ ⓑ ⓒ ⓓ ⓔ
19. ⓐ ⓑ ⓒ ⓓ ⓔ
20. ⓐ ⓑ ⓒ ⓓ ⓔ

► Section 2

1. ⓐ ⓑ ⓒ ⓓ ⓔ
2. ⓐ ⓑ ⓒ ⓓ ⓔ
3. ⓐ ⓑ ⓒ ⓓ ⓔ
4. ⓐ ⓑ ⓒ ⓓ ⓔ
5. ⓐ ⓑ ⓒ ⓓ ⓔ
6. ⓐ ⓑ ⓒ ⓓ ⓔ
7. ⓐ ⓑ ⓒ ⓓ ⓔ
8. ⓐ ⓑ ⓒ ⓓ ⓔ

9.

	/	/	
.	.	.	.
	0	0	0
1	1	1	1
2	2	2	2
3	3	3	3
4	4	4	4
5	5	5	5
6	6	6	6
7	7	7	7
8	8	8	8
9	9	9	9

10.

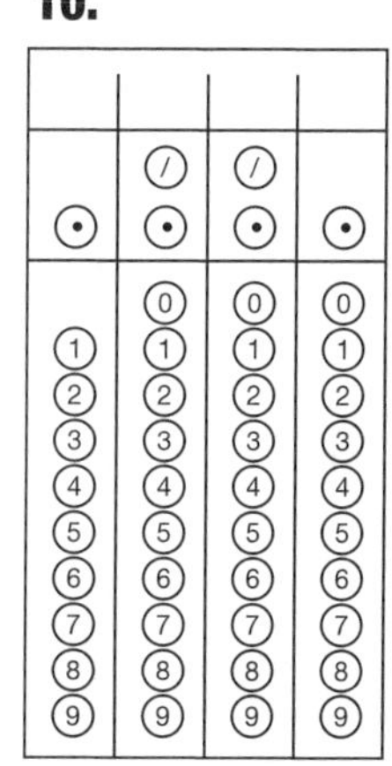

11.

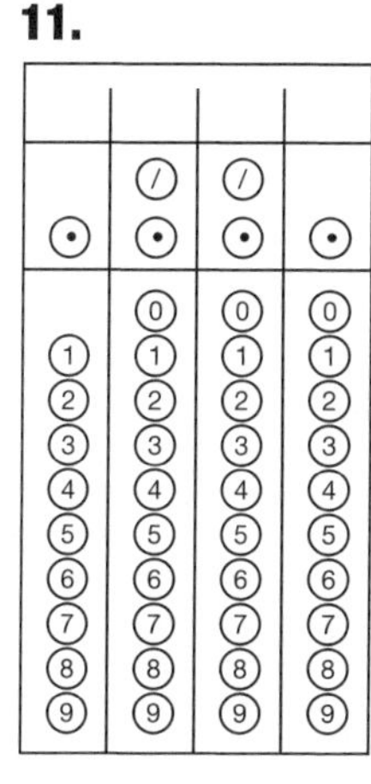

12.

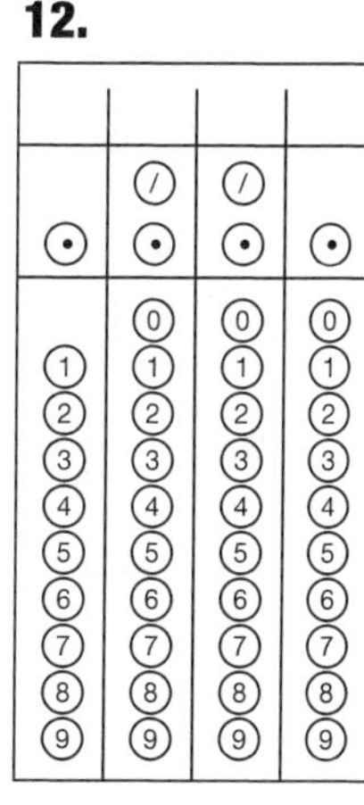

13.

	/	/	
.	.	.	.
	0	0	0
1	1	1	1
2	2	2	2
3	3	3	3
4	4	4	4
5	5	5	5
6	6	6	6
7	7	7	7
8	8	8	8
9	9	9	9

14.

	/	/	
.	.	.	.
	0	0	0
1	1	1	1
2	2	2	2
3	3	3	3
4	4	4	4
5	5	5	5
6	6	6	6
7	7	7	7
8	8	8	8
9	9	9	9

15.

	/	/	
.	.	.	.
	0	0	0
1	1	1	1
2	2	2	2
3	3	3	3
4	4	4	4
5	5	5	5
6	6	6	6
7	7	7	7
8	8	8	8
9	9	9	9

16.

	/	/	
.	.	.	.
	0	0	0
1	1	1	1
2	2	2	2
3	3	3	3
4	4	4	4
5	5	5	5
6	6	6	6
7	7	7	7
8	8	8	8
9	9	9	9

17.

	/	/	
.	.	.	.
	0	0	0
1	1	1	1
2	2	2	2
3	3	3	3
4	4	4	4
5	5	5	5
6	6	6	6
7	7	7	7
8	8	8	8
9	9	9	9

18.

	/	/	
.	.	.	.
	0	0	0
1	1	1	1
2	2	2	2
3	3	3	3
4	4	4	4
5	5	5	5
6	6	6	6
7	7	7	7
8	8	8	8
9	9	9	9

▶ Section 3

1. ⓐ ⓑ ⓒ ⓓ ⓔ
2. ⓐ ⓑ ⓒ ⓓ ⓔ
3. ⓐ ⓑ ⓒ ⓓ ⓔ
4. ⓐ ⓑ ⓒ ⓓ ⓔ
5. ⓐ ⓑ ⓒ ⓓ ⓔ
6. ⓐ ⓑ ⓒ ⓓ ⓔ
7. ⓐ ⓑ ⓒ ⓓ ⓔ
8. ⓐ ⓑ ⓒ ⓓ ⓔ
9. ⓐ ⓑ ⓒ ⓓ ⓔ
10. ⓐ ⓑ ⓒ ⓓ ⓔ
11. ⓐ ⓑ ⓒ ⓓ ⓔ
12. ⓐ ⓑ ⓒ ⓓ ⓔ
13. ⓐ ⓑ ⓒ ⓓ ⓔ
14. ⓐ ⓑ ⓒ ⓓ ⓔ
15. ⓐ ⓑ ⓒ ⓓ ⓔ
16. ⓐ ⓑ ⓒ ⓓ ⓔ

▶ Section 1

1. If $m = 6$, then the expression $\frac{m^2}{3} - 4m + 10$ is equal to
- **a.** –12.
- **b.** –2.
- **c.** 6.
- **d.** 12.
- **e.** 22.

2. Which of the following is the midpoint of a line with endpoints at (–2,–8) and (8,0)?
- **a.** (3,4)
- **b.** (3,–4)
- **c.** (–5,4)
- **d.** (5,–4)
- **e.** (6,–8)

3. If $4x + 5 = 15$, then $10x + 5 =$
- **a.** 2.5.
- **b.** 15.
- **c.** 22.5.
- **d.** 25.
- **e.** 30.

4. A music store offers customized guitars. A buyer has four choices for the neck of the guitar, two choices for the body of the guitar, and six choices for the color of the guitar. The music store offers
- **a.** 12 different guitars.
- **b.** 16 different guitars.
- **c.** 24 different guitars.
- **d.** 36 different guitars.
- **e.** 48 different guitars.

5. Which of the following is the set of positive factors of 12 that are NOT multiples of 2?
- **a.** { }
- **b.** {1}
- **c.** {1, 3}
- **d.** {1, 2, 3}
- **e.** {2, 4, 6, 12}

6.

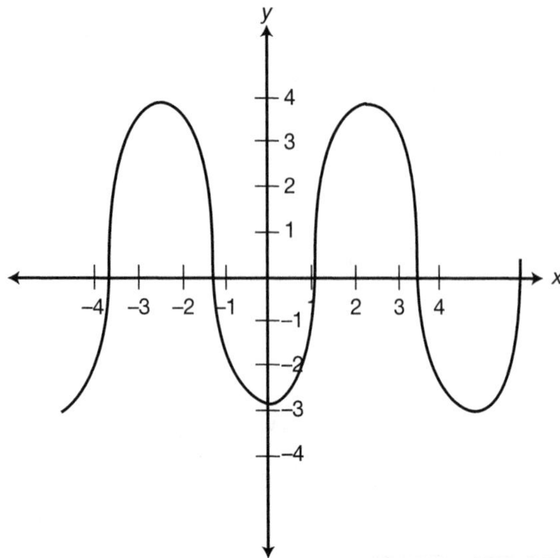

The graph of $f(x)$ is shown above. How many values can be found for $f(3)$?

a. 0
b. 1
c. 2
d. 4
e. cannot be determined

7. The expression $\frac{x^2 + 5x}{x^3 - 25x}$ can be reduced to

a. 1.
b. $\frac{5}{x^2 - 25}$.
c. $x + 5$.
d. $\frac{1}{x - 5}$.
e. $\frac{x}{x + 5}$.

8. Which of the following is the vertex of the parabola which is the graph of the equation $y = (x + 1)^2 + 2$?

a. (–1,–2)
b. (1,–2)
c. (–1,2)
d. (1,2)
e. (2,–1)

9. $a^{\frac{b}{c}}$ is equivalent to

a. $\sqrt[c]{a^b}$.

b. $\sqrt[b]{a^c}$.

c. $\frac{1}{a^{\frac{c}{b}}}$.

d. $\frac{\sqrt{a^b}}{c}$

e. $\frac{a^b}{c}$.

10. If the statement "No penguins live at the North Pole" is true, which of the following statements must also be true?

a. All penguins live at the South Pole.
b. If Flipper is not a penguin, then he lives at the North Pole.
c. If Flipper is not a penguin, then he does not live at the North Pole.
d. If Flipper does not live at the North Pole, then he is a penguin.
e. If Flipper lives at the North Pole, then he is not a penguin.

11. If $p < 0$, $q > 0$, and $r > p$, then which of the following must be true?

a. $p + r > 0$
b. $r^p < r^q$
c. $pr < rq$
d. $r + q > q$
e. $p + r < r + q$

12.

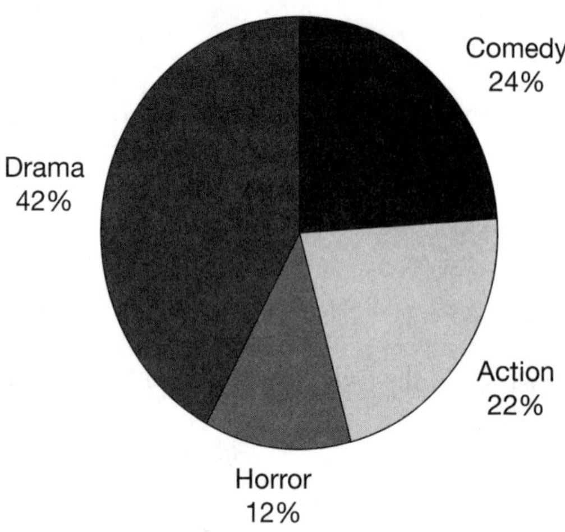

The pie chart above shows the distribution of video rentals from Al's Video Vault for a single night. If 250 videos were rented that night, how many more action movies were rented than horror movies?

a. 10
b. 20
c. 22
d. 25
e. 30

13.

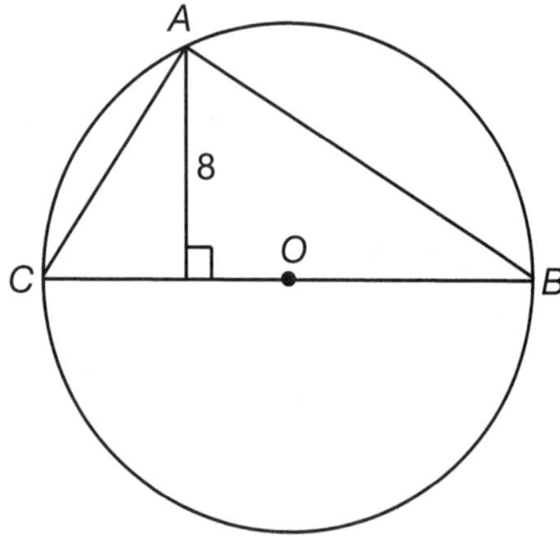

If the circumference of the circle in the diagram above is 20π units, what is the area of triangle *ABC*?

a. 40 square units
b. 80 square units
c. 80π square units
d. 160 square units
e. 160π square units

14. The area of an isosceles right triangle is 18 cm^2. What is the length of the hypotenuse of the triangle?

a. 6 cm
b. $6\sqrt{2}$ cm
c. $18\sqrt{2}$ cm
d. $18\sqrt{3}$ cm
e. $36\sqrt{2}$ cm

15. If $a < \frac{43}{3x} < b$, and $a = 4$ and $b = 8$, which of the following could be true?

a. $x < a$
b. $x > b$
c. $a < x < b$
d. $4 < x < 8$
e. none of the above

16. The length of a rectangle is one greater than three times its width. If the perimeter of the rectangle is 26 feet, what is the area of the rectangle?

a. 13 ft^2
b. 24 ft^2
c. 30 ft^2
d. 78 ft^2
e. 100 ft^2

17.

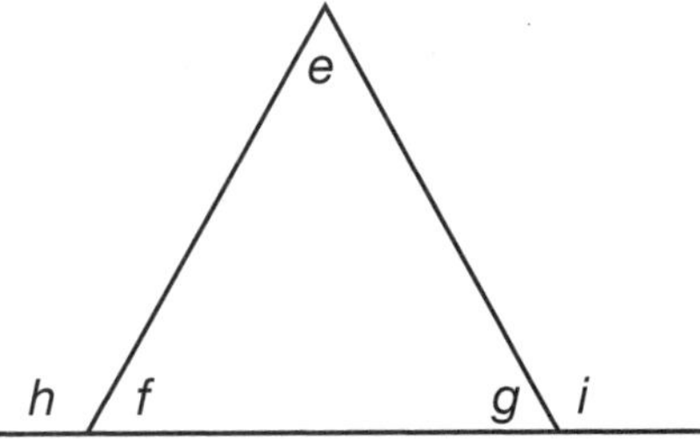

Based on the diagram above, which of the following is true?

a. $i = e + f$
b. $g + i = h + e$
c. $e + i = e + h$
d. $e + g + i = 180$
e. $e + f + g + h + i = 360$

18. Which of the following is an irrational number?

a. $\sqrt{\frac{4}{9}}$

b. 4^{-3}

c. $-(\sqrt{3}\sqrt{3})$

d. $\frac{\sqrt{72}}{\sqrt{200}}$

e. $(\sqrt{32})^3$

19.

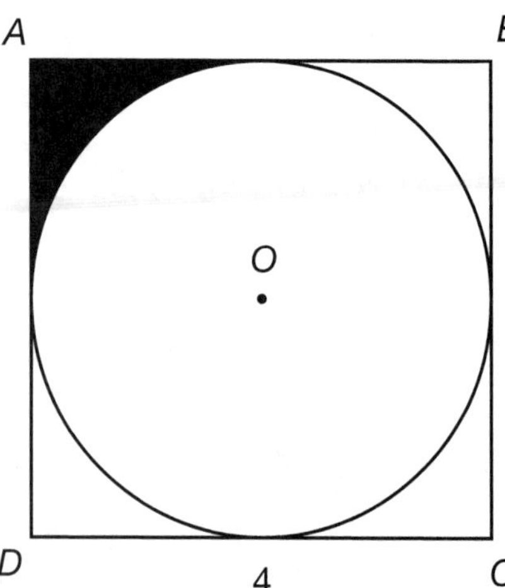

In the diagram above, the length of a side of square *ABCD* is four units. What is the area of the shaded region?

a. 4
b. $4 - \pi$
c. $4 - 4\pi$
d. 16π
e. $16 - 4\pi$

20. The value of *d* is increased 50%, then decreased 50%. Compared to its original value, the value of *d* is now

a. 25% smaller.
b. 25% larger.
c. 50% smaller.
d. 50% larger.
e. the same.

▶ Section 2

1. Which of the following expressions is undefined when $x = -2$?

a. $y = \frac{x+2}{x-2}$

b. $y = \frac{x^2+4x+4}{x}$

c. $y = \frac{2x+4}{x^2-4x+4}$

d. $y = \frac{x^2+3x+2}{-x^2+2}$

e. $y = \frac{x^2+2x+2}{x^2+6x+8}$

2. If graphed, which of the following pairs of equations would be parallel to each other?

a. $y = 2x + 4$, $y = x + 4$
b. $y = 3x + 3$, $y = -\frac{1}{3}x - 3$
c. $y = 4x + 1$, $y = -4x + 1$
d. $y = 5x + 5$, $y = \frac{1}{5}x + 5$
e. $y = 6x + 6$, $y = 6x - 6$

3. If $\frac{a}{b-4} = \frac{4b}{a} + 1$, then when $a = 8$, b could be equal to

a. –2.
b. 4.
c. 6.
d. 7.
e. 8.

4. The average of five consecutive odd integers is –21. What is the least of these integers?

a. –17
b. –19
c. –21
d. –23
e. –25

5. Line *AC* is a diagonal of square *ABCD*. What is the sine of angle *ACB*?

a. $\frac{1}{2}$

b. $\sqrt{2}$

c. $\frac{\sqrt{2}}{2}$

d. $\frac{\sqrt{3}}{2}$

e. cannot be determined

6. If the height of a cylinder is doubled and the radius of the cylinder is halved, the volume of the cylinder

a. remains the same.
b. becomes twice as large.
c. becomes half as large.
d. becomes four times larger.
e. becomes four times smaller.

7. $\dfrac{\frac{b}{a} - a}{\frac{1}{a^{-1}}} =$

a. b

b. $b - a^2$

c. $\frac{b}{a} - 1$

d. $\frac{b}{a^2} - 1$

e. $\frac{b}{a^2} - a$

8. The ratio of the number of cubic units in the volume of a cube to the number of square units in the surface area of the cube is 2:3. What is the surface area of the cube?

a. 16 square units
b. 24 square units
c. 64 square units
d. 96 square units
e. 144 square units

9. If a number is chosen at random from a set that contains only the whole number factors of 24, what is the probability that the number is either a multiple of four or a multiple of six?

10. There are 750 students in the auditorium for an assembly. When the assembly ends, the students begin to leave. If 32% of the students have left so far, how many students are still in the auditorium?

11. If point A is at $(-1,2)$ and point B is at $(11,-7)$, what is length of line AB?

12. Robert is practicing for the long jump competition. His first four jumps measure 12.4 ft, 18.9 ft, 17.3 ft, and 15.3 ft, respectively. If he averages 16.3 feet for his first five jumps, what is the length in feet of his fifth jump?

13. There are seven students on the trivia team. Mr. Randall must choose four students to participate in the trivia challenge. How many different groups of four students can Mr. Randall form?

14.

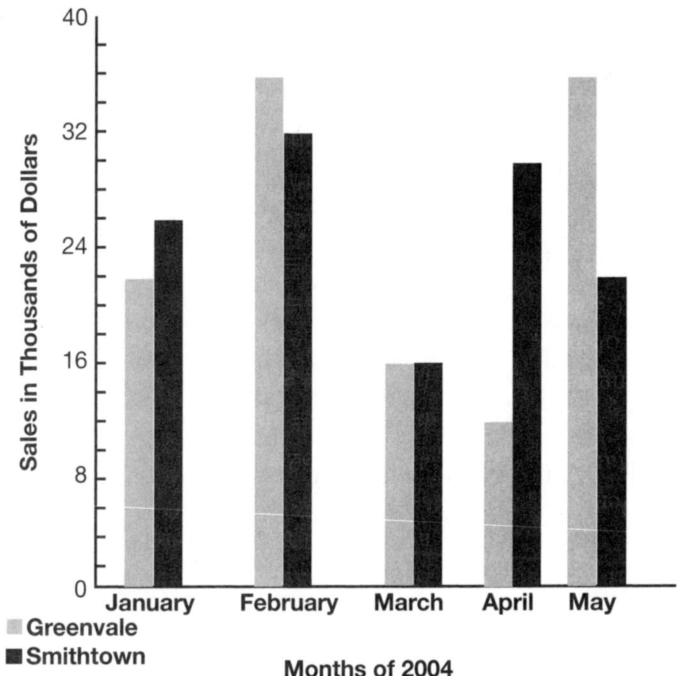

The graph above shows the sales by month for the Greenvale and Smithtown branches of SuperBooks. From January through May, how much more money did the Smithtown branch gross in sales than the Greenvale branch?

15.

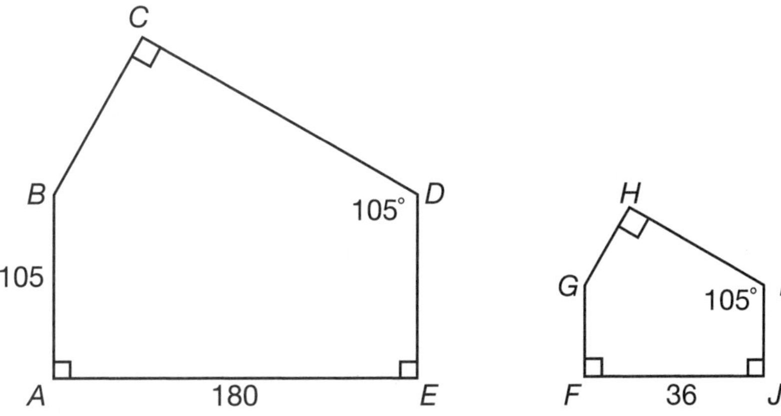

In the diagram above, what is the length of side *FG*?

16. DeDe and Mike both run the length of a two-mile field. If DeDe runs 5 mph and Mike runs 6 mph, how many more minutes does it take DeDe to run the field?

17. Point A of rectangle $ABCD$ is located at $(-3,12)$ and point C is located at $(9,5)$. What is the area of rectangle $ABCD$?

18.

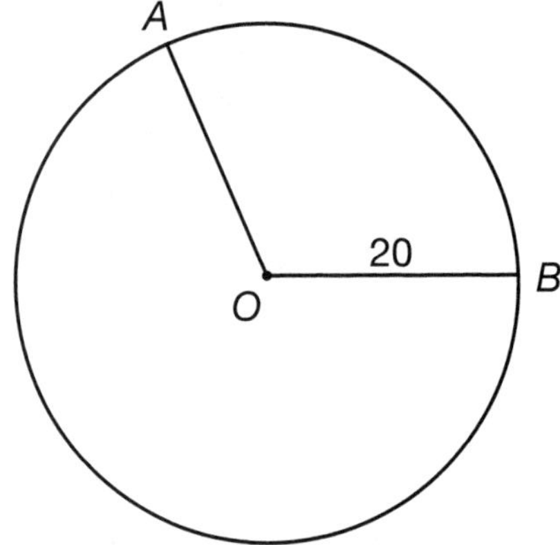

In the diagram above, the radius of the circle is 20 units and the length of arc AB is 15π units. What is the measure in degrees of angle AOB?

► Section 3

1. All of the following are less than $\frac{2}{5}$ EXCEPT

a. $\frac{1}{3}$.

b. 0.04.

c. $\frac{3}{8}$.

d. $\frac{3}{7}$.

e. 0.0404.

2. If $3x - y = 2$ and $2y - 3x = 8$, which of the following is equal to $\frac{x}{y}$?

a. $\frac{2}{3}$

b. $\frac{2}{5}$

c. $2\frac{1}{2}$

d. 4

e. 6

3. Which of the following sets of numbers contains all and only the roots of the equation $f(x) = x^3 + 7x^2 - 8x$?

a. $\{-8, 1\}$
b. $\{8, -1\}$
c. $\{0, -8, 1\}$
d. $\{0, 8, -1\}$
e. $\{0, -1, -8, 1, 8\}$

4. What is the equation of the line that passes through the points (2,3) and (–2,5)?

a. $y = x + 1$
b. $y = -\frac{1}{2}x + 4$
c. $y = -\frac{1}{2}x$
d. $y = -\frac{3}{2}x$
e. $y = -\frac{3}{2}x + 2$

5. An empty crate weighs 8.16 kg and an orange weighs 220 g. If Jon can lift 11,000 g, how many oranges can he pack in the crate before lifting it onto his truck?

a. 12
b. 13
c. 37
d. 46
e. 50

6. The measures of the length, width, and height of a rectangular prism are in the ratio 2:6:5. If the volume of the prism is 1,620 mm^3, what is the width of the prism?

a. 3 mm
b. 6 mm
c. 9 mm
d. 18 mm
e. 27 mm

7. A box contains five blue pens, three black pens, and two red pens. If every time a pen is selected, it is removed from the box, what is the probability of selecting a black pen followed by a blue pen?

a. $\frac{1}{6}$
b. $\frac{1}{10}$
c. $\frac{1}{50}$
d. $\frac{3}{20}$
e. $\frac{77}{90}$

8.

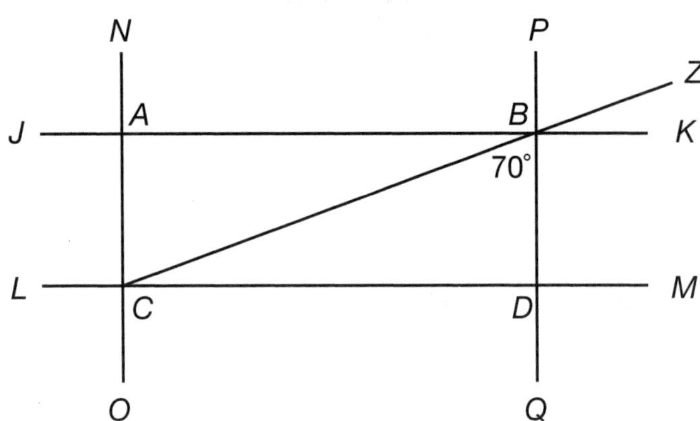

In the diagram above, lines *NO* and *PQ* are parallel to each other and perpendicular to lines *JK* and *LM*. Line *JK* is parallel to line *LM*. If angle *CBD* is 70 degrees, what is the measure of angle *ZBK*?

a. 10 degrees
b. 20 degrees
c. 70 degrees
d. 90 degrees
e. 110 degrees

9. Monica sells pretzels in the cafeteria every school day for a week. She sells 14 pretzels on Monday, 12 pretzels on Tuesday, 16 pretzels on Wednesday, and 12 pretzels on Thursday. Then, she calculates the mean, median, and mode of her sales. If she sells 13 pretzels on Friday, then

a. the mode will increase.
b. the mean will stay the same.
c. the median will stay the same.
d. the median will decrease.
e. the mean will increase.

10. What is the tenth term of the pattern below?

$\frac{10}{1{,}024}, \frac{9}{512}, \frac{8}{256}, \frac{7}{128}, \ldots$

a. $\frac{1}{2}$
b. $\frac{2}{9}$
c. $\frac{9}{2}$
d. $\frac{9}{4}$
e. 1

11. Which of the following statements is always true if p is a rational number?

a. $|p| < |3p|$
b. $|p^2| > |p + 1|$
c. $|-p| > p$
d. $|p^3| > |p^2|$
e. $|p^{-p}| > p^{-p}$

12.

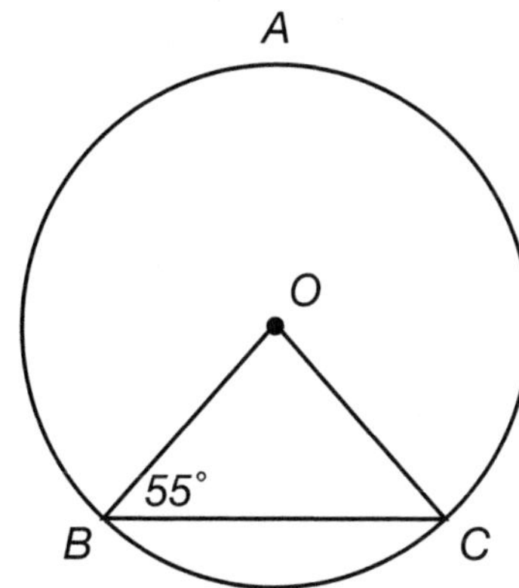

In the diagram above, side $OB \cong$ side OC. Which of the following is the measure of minor arc BC?

a. 27.5 degrees
b. 45 degrees
c. 55 degrees
d. 70 degrees
e. 110 degrees

13. If $g \wedge h = \frac{2h}{g}$, then $(h \wedge g) \wedge h =$

a. $2h$.
b. $4h$.
c. $\frac{h^2}{g}$.
d. $\frac{2h^2}{g}$.
e. $\frac{4h^2}{g}$.

14. Four copy machines make 240 total copies in three minutes. How long will it take five copy machines to make the same number of copies?

a. 2 minutes
b. 2 minutes, 15 seconds
c. 2 minutes, 24 seconds
d. 2 minutes, 45 seconds
e. 3 minutes, 36 seconds

15. If 40% of j is equal to 50% of k, then j is

a. 10% larger than k.
b. 15% larger than k.
c. 20% larger than k.
d. 25% larger than k.
e. 80% larger than k.

16.

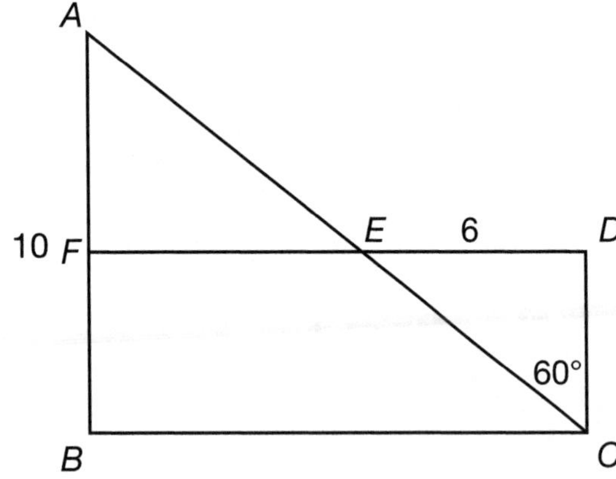

In the diagram above, *FDCB* is a rectangle. Line *ED* is six units long, line *AB* is ten units long, and the measure of angle *ECD* is 60 degrees. What is the length of line *AE*?

a. 8
b. $\frac{\sqrt{3}}{2}$
c. 20
d. $20 - \frac{\sqrt{3}}{2}$
e. $20 - 4\sqrt{3}$

▶ Answer Key

Section 1 Answers

1. b. Substitute 6 for m: $\frac{6^2}{3} - 4(6) + 10 = \frac{36}{3} - 24 + 10 = 12 - 14 = -2$.

2. b. The midpoint of a line is equal to the average of the x- and y-coordinates of its endpoints. The average of the x-coordinates $= \frac{-2+8}{2} = \frac{6}{2} = 3$. The average of the y-coordinates $= \frac{-8+0}{2} = -\frac{8}{2} = -4$. The midpoint of this line is at (3,–4).

3. e. If $4x + 5 = 15$, then $4x = 10$ and $x = 2.5$. Substitute 2.5 for x in the second equation: $10(2.5) + 5 = 25 + 5 = 30$.

4. e. To find the total number of different guitars that are offered, multiply the number of neck choices by the number of body choices by the number of color choices: $(4)(2)(6) = 48$ different guitars.

5. c. The set of positive factors of 12 is {1, 2, 3, 4, 6, 12}. All of the even numbers (2, 4, 6, and 12) are multiples of 2. The only positive factors of 12 that are not multiples of 2 are 1 and 3.

6. b. Be careful—the question asks you for the number of values of $f(3)$, not $f(x) = 3$. In other words, how many y values can be generated when $x = 3$? If the line $x = 3$ is drawn on the graph, it passes through only one point. There is only one value for $f(3)$.

7. d. Factor the numerator and denominator of the fraction:
$(x^2 + 5x) = x(x + 5)$
$(x^3 - 25x) = x(x + 5)(x - 5)$
There is an x term and an $(x + 5)$ term in both the numerator and denominator. Cancel those terms, leaving the fraction $\frac{1}{x-5}$.

8. c. The equation of a parabola with its turning point c units to the left of the y-axis is written as $y = (x + c)^2$. The equation of a parabola with its turning point d units above the x-axis is written as $y = x^2 + d$. The vertex of the parabola formed by the equation $y = (x + 1)^2 + 2$ is found one unit to the left of the y-axis and two units above the x-axis, at the point (–1,2). Alternatively, test each answer choice by plugging the x value of the choice into the equation and solving for y. Only the coordinates in choice **c**, (–1, 2), represent a point on the parabola ($y = (x + 1)^2 + 2$, $2 = (-1 + 1)^2 + 2$, $2 = 0^2 + 2$, $2 = 2$), so it is the only point of the choices given that could be the vertex of the parabola.

9. a. When a base is raised to a fractional exponent, raise the base to the power given by the numerator and take the root given by the denominator. Raise the base, a, to the bth power, since b is the numerator of the exponent. Then, take the cth rooth of that: $\sqrt[c]{a^b}$.

10. e. No penguins live at the North Pole, so anything that lives at the North Pole must not be a penguin. If Flipper lives at the North Pole, then he, like all things at the North Pole, is not a penguin.

11. e. If $p < 0$ and $q > 0$, then $p < q$. Since $p < q$, p plus any value will be less than q plus that same value (whether positive or negative). Therefore, $p + r < r + q$.

12. d. 22% of the movies rented were action movies; $250(0.22) = 55$ movies; 12% of the movies rented were horror movies; $250(0.12) = 30$ movies. There were $55 - 30 = 25$ more action movies rented than horror movies.

13. b. The circumference of a circle is equal to $2\pi r$, where r is the radius of the circle. If the circumference of the circle $= 20\pi$ units, then the radius of the circle is equal to ten units. The base of triangle ABC is the diameter of the circle, which is twice the radius. The base of the triangle is 20 units and the height of the triangle is eight units. The area of a triangle is equal to $\frac{1}{2}bh$, where b is the base of the triangle and h is the height of the triangle. The area of triangle $ABC = \frac{1}{2}(8)(20) = \frac{1}{2}(160) = 80$ square units.

14. b. The area of a triangle is equal to $\frac{1}{2}bh$, where b is the base of the triangle and h is the height of the triangle. The base and height of an isosceles right triangle are equal in length. Therefore, $\frac{1}{2}b^2 = 18$, $b^2 = 36$, $b = 6$. The legs of the triangle are 6 cm. The hypotenuse of an isosceles right triangle is equal to the length of one leg multiplied by $\sqrt{2}$. The hypotenuse of this triangle is equal to $6\sqrt{2}$ cm.

15. a. If $a = 4$, x could be less than a. For example, x could be 3: $4 < \frac{43}{3(3)} < 8$, $4 < \frac{43}{9} < 8$, $4 < 4\frac{7}{9} < 8$. Although $x < a$ is not true for all values of x, it is true for some values of x.

16. c. The perimeter of a rectangle is equal to $2l + 2w$, where l is the length of the rectangle and w is the width of the rectangle. If the length is one greater than three times the width, then set the width equal to x and set the length equal to $3x + 1$:

$2(3x + 1) + 2(x) = 26$

$6x + 2 + 2x = 26$

$8x = 24$

$x = 3$

The width of the rectangle is 3 ft and the length of the rectangle is 10 ft. The area of a rectangle is equal to lw; $(10 \text{ ft})(3 \text{ ft}) = 30 \text{ ft}^2$.

17. a. The measure of an exterior angle of a triangle is equal to the sum of the two interior angles of the triangle to which the exterior angle is NOT supplementary. Angle i is supplementary to angle g, so the sum of the interior angles e and f is equal to the measure of angle i: $i = e + f$.

18. e. An irrational number is a number that cannot be expressed as a repeating or terminating decimal. $(\sqrt{32})^3 = (\sqrt{32})(\sqrt{32})(\sqrt{32}) = 32\sqrt{32} = 32\sqrt{16}\sqrt{2} = (32)(4)\sqrt{2} = 128\sqrt{2}$. $\sqrt{2}$ cannot be expressed as a repeating or terminating decimal, therefore, $128\sqrt{2}$ is an irrational number.

19. b. The area of a square is equal to s^2, where s is the length of a side of the square. The area of $ABCD$ is $4^2 = 16$ square units. The area of a circle is equal to πr^2, where r is the radius of the circle. The diameter of the circle is four units. The radius of the circle is $\frac{4}{2}$ = two square units. The area of the circle is equal to $\pi(2)^2 = 4\pi$. The shaded area is equal to one-fourth of the difference between the area of the square and the area of the circle: $\frac{1}{4}(16 - 4\pi) = 4 - \pi$.

20. a. To increase d by 50%, multiply d by 1.5: $d = 1.5d$. To find 50% of $1.5d$, multiply $1.5d$ by 0.5: $(1.5d)(0.5) = 0.75d$. Compared to its original value, d is now 75% of what it was. The value of d is now 25% smaller.

Section 2 Answers

1. e. An expression is undefined when a denominator of the expression is equal to zero. When $x = -2$, $x^2 + 6x + 8 = (-2)^2 + 6(-2) + 8 = 4 - 12 + 8 = 0$.

2. e. Parallel lines have the same slope. The lines $y = 6x + 6$ and $y = 6x - 6$ both have a slope of 6, so they are parallel to each other.

3. c. Substitute 8 for a: $\frac{8}{b-4} = \frac{4b}{8} + 1$. Rewrite 1 as $\frac{8}{8}$ and add it to $\frac{4b}{8}$, then cross multiply:

$\frac{8}{b-4} = \frac{4b+8}{8}$

$4b^2 - 8b - 32 = 64$

$b^2 - 2b - 8 = 16$

$b^2 - 2b - 24 = 0$

$(b - 6)(b + 4) = 0$

$b - 6 = 0, b = 6$

$b + 4 = 0, b = -4$

4. e. If the average of five consecutive odd integers is −21, then the third integer must be −21. The two larger integers are −19 and −17 and the two lesser integers are −23 and −25. −25 is the least of the five integers. Remember, the more a number is negative, the less is its value.

5. c. A square has four right (90-degree) angles. The diagonals of a square bisect its angles. Diagonal AC bisects C, forming two 45-degree angles, angle ACB and angle ACD. The sine of 45 degrees is equal to $\frac{\sqrt{2}}{2}$.

6. c. The volume of a cylinder is equal to πr^2h, where r is the radius of the cylinder and h is the height. The volume of a cylinder with a radius of 1 and a height of 1 is π. If the height is doubled and the radius is halved, then the volume becomes $\pi(\frac{1}{2})^2(2)(1) = \pi(\frac{1}{4})2 = \frac{1}{2}\pi$. The volume of the cylinder has become half as large.

7. d. $\frac{1}{a^{-1}} = \frac{1}{\frac{1}{a}} = a$, $\frac{\frac{b}{a} - a}{a} = (\frac{b}{a} - a)(\frac{1}{a}) = \frac{b}{a^2 - 1}$

8. d. The volume of a cube is equal to e^3, where e is the length of an edge of the cube. The surface area of a cube is equal to $6e^2$. If the ratio of the number of cubic units in the volume to the number of square units in the surface area is 2:3, then three times the volume is equal to two times the surface area:

$3e^3 = 2(6e^2)$

$3e^3 = 12e^2$

$3e = 12$

$e = 4$

The edge of the cube is four units and the surface area of the cube is $6(4)^2 = 96$ square units.

9. $\frac{5}{8}$ The set of whole number factors of 24 is {1, 2, 3, 4, 6, 8, 12, 24}. Of these numbers, four (4, 8, 12, 24) are multiples of four and three (6, 12, 24) are multiples of six. Be sure not to count 12 and 24 twice—there are five numbers out of the eight factors of 24 that are a multiple of either four or six. Therefore, the probability of selecting one of these numbers is $\frac{5}{8}$.

10. 510 If 32% of the students have left the auditorium, then 100 – 32 = 68% of the students are still in the auditorium; 68% of 750 = (0.68)(750) = 510 students.

11. 15 Use the distance formula to find the distance from (–1,2) to (11,–7):

Distance = $\sqrt{(x_2 - x_1)^2 + (y_2 - y_1)^2}$

Distance = $\sqrt{(11 - (-1))^2 + ((-7) - 2)^2}$

Distance = $\sqrt{(12)^2 + (-9)^2}$

Distance = $\sqrt{144 + 81}$

Distance = $\sqrt{225}$

Distance = 15 units

12. 17.6 If Robert averages 16.3 feet for five jumps, then he jumps a total of (16.3)(5) = 81.5 feet. The sum of Robert's first four jumps is 12.4 ft + 18.9 ft + 17.3 ft + 15.3 ft = 63.9 ft. Therefore, the measure of his fifth jump is equal to 81.5 ft – 63.9 ft = 17.6 ft.

13. 35 The order of the four students chosen does not matter. This is a "seven-choose-four" combination problem—be sure to divide to avoid counting duplicates: $\frac{(7)(6)(5)(4)}{(4)(3)(2)(1)} = \frac{840}{24} =$ 35. There are 35 different groups of four students that Mr. Randall could form.

14. 4,000 The Greenvale sales, represented by the light bars, for the months of January through May respectively were \$22,000, \$36,000, \$16,000, \$12,000, and \$36,000, for a total of \$122,000. The Smithtown sales, represented by the dark bars, for the months of January through May respectively were \$26,000, \$32,000, \$16,000, \$30,000, and \$22,000, for a total of \$126,000. The Smithtown branch grossed \$126,000 – \$122,000 = \$4,000 more than the Greenvale branch.

15. 21 Both figures contain five angles. Each figure contains three right angles and an angle labeled 105 degrees. Therefore, the corresponding angles in each figure whose measures are not given (angles B and G, respectively) must also be equal, which makes the two figures similar. The lengths of the sides of similar figures are in the same ratio. The length of side FJ is 36 units and the length of its corresponding side, AE, in figure $ABCDE$ is 180 units. Therefore, the ratio of side FJ to side AE is 36:180 or 1:5. The lengths of sides FG and AB are in the same ratio. If the length of side FG is x, then: $\frac{x}{105} = \frac{1}{5}$, $5x = 105$, $x = 21$. The length of side FG is 21 units.

16. 4 DeDe runs 5 mph, or 5 miles in 60 minutes. Use a proportion to find how long it would take for DeDe to run 2 miles: $\frac{5}{60} = \frac{2}{x}$, $5x = 120$, $x = 24$ minutes. Greg runs 6 mph, or 6 miles in 60 minutes. Therefore, he runs 2 miles in

$\frac{6}{60} = \frac{2}{x}$, $6x = 120$, $x = 20$ minutes. It takes DeDe $24 - 20 = 4$ minutes longer to run the field.

17. 84 If point A is located at (–3,12) and point C is located at (9,5), that means that either point B or point D has the coordinates (–3,5) and the other has the coordinates (9,12). The difference between the different x values is $9 - (-3) = 12$ and the difference between the different y values is $12 - 5 = 7$. The length of the rectangle is 12 units and the width of the rectangle is seven units. The area of a rectangle is equal to its length multiplied by its width, so the area of $ABCD = (12)(7) = 84$ square units.

18. 135 The length of an arc is equal to the circumference of the circle multiplied by the measure of the angle that intercepts the arc divided by 360. The arc measures 15π units, the circumference of a circle is 2π multiplied by the radius, and the radius of the circle is 20 units. If x represents the measure of angle AOB, then:

$15\pi = \frac{x}{360}2\pi(20)$

$15 = \frac{x}{360}(40)$

$15 = \frac{x}{9}$

$x = 135$

The measure of angle AOB is 135 degrees.

Section 3 Answers

1. d. $\frac{2}{5} = 0.40$. $\frac{3}{7} \approx 0.43$. Comparing the hundredths digits, $3 > 0$, therefore, $0.43 > 0.40$ and $\frac{3}{7} > \frac{2}{5}$.

2. b. Solve $3x - y = 2$ for y: $-y = -3x + 2$, $y = 3x - 2$. Substitute $3x - 2$ for y in the second equation and solve for x:

$2(3x - 2) - 3x = 8$

$6x - 4 - 3x = 8$

$3x - 4 = 8$

$3x = 12$

$x = 4$

Substitute the value of x into the first equation to find the value of y:

$3(4) - y = 2$

$12 - y = 2$

$y = 10$

$\frac{x}{y} = \frac{4}{10} = \frac{2}{5}$.

3. c. The roots of an equation are the values for which the equation evaluates to zero. Factor $x^3 + 7x^2 - 8x$: $x^3 + 7x^2 - 8x = x(x^2 + 7x - 8) = x(x + 8)(x - 1)$. When $x = 0, -8$, or 1, the equation $f(x) = x^3 + 7x^2 - 8x$ is equal to zero. The set of roots is {0, –8, 1}.

4. b. First, find the slope of the line. The slope of a line is equal to the change in y values divided by the change in x values of two points on the line. The y value increases by 2 (5 – 3) and the x value decreases by 4 (–2 – 2). Therefore, the slope of the line is equal to $-\frac{2}{4}$, or $-\frac{1}{2}$. The equation of the line is $y = -\frac{1}{2}x + b$, where b is the y-intercept. Use either of the two given points to solve for b:

$3 = -\frac{1}{2}(2) + b$

$3 = -1 + b$

$b = 4$

The equation of the line that passes through the points (2,3) and (–2,5) is $y = -\frac{1}{2}x + 4$.

5. a. The empty crate weighs 8.16 kg, or 8,160 g. If Jon can lift 11,000 g and one orange weighs 220 g, then the number of oranges that he can pack into the crate is equal to $\frac{11{,}000 - 8{,}160}{220} = \frac{2{,}840}{220} \approx 12.9$. Jon cannot pack a fraction of an orange. He can pack 12 whole oranges into the crate.

6. d. The volume of a prism is equal to lwh, where l is the length of the prism, w is the width of the prism, and h is the height of the prism:

$(2x)(6x)(5x) = 1{,}620$

$60x^3 = 1{,}620$

$x^3 = 27$

$x = 3$

The length of the prism is $2(3) = 6$ mm, the width of the prism is $6(3) = 18$ mm, and the height of the prism is $5(3) = 15$ mm.

7. a. At the start, there are 5 + 3 + 2 = 10 pens in the box, 3 of which are black. Therefore, the probability of selecting a black pen is $\frac{3}{10}$. After the black pen is removed, there are nine pens remaining in the box, five of which are blue. The probability of selecting a blue pen second is $\frac{5}{9}$. To find the probability that both events will happen, multiply the probability of the first event by the probability of the second event: $(\frac{3}{10})(\frac{5}{9}) = \frac{15}{90} = \frac{1}{6}$.

8. b. Angle *CBD* and angle *PBZ* are alternating angles—their measures are equal. Angle *PBZ* = 70 degrees. Angle *PBZ* + angle *ZBK* form angle *PBK*. Line *PQ* is perpendicular to line *JK*; therefore, angle *PBK* is a right angle (90 degrees). Angle *ZBK* = angle *PBK* – angle *PBZ* = 90 – 70 = 20 degrees.

9. c. For the first four days of the week, Monica sells 12 pretzels, 12 pretzels, 14 pretzels, and 16 pretzels. The median value is the average of the second and third values: $\frac{12+14}{2} = \frac{26}{2} = 13$. If Monica sells 13 pretzels on Friday, the median will still be 13. She will have sold 12 pretzels, 12 pretzels, 13 pretzels, 14 pretzels, and 16 pretzels. The median stays the same.

10. a. The denominator of each term in the pattern is equal to 2 raised to the power given in the numerator. The numerator decreases by 1 from one term to the next. Since 10 is the numerator of the first term, 10 – 9, or 1, will be the numerator of the tenth term. $2^1 = 2$, so the tenth term will be $\frac{1}{2}$.

11. a. No matter whether *p* is positive or negative, or whether *p* is a fraction, whole number, or mixed number, the absolute value of three times any number will always be positive and greater than the absolute value of that number.

12. d. Line $OB \cong$ line *OC*, which means the angles opposite line *OB* and *OC* (angles *C* and *B*) are congruent. Since angle *B* = 55 degrees, then angle *C* = 55 degrees. There are 180 degrees in a triangle, so the measure of angle *O* is equal to 180 – (55 + 55) = 180 – 110 = 70 degrees. Angle *O* is a central angle. The measure of its intercepted arc, minor arc *BC*, is equal to the measure of angle *O*, 70 degrees.

13. c. This uses the same principles as #10 in Test 1, section 2. ^ is a function definition just as # was a function definition. ^ means "take the value after the ^ symbol, multiply it by 2, and divide it by the value before the ^ symbol." So, *h*^*g* is equal to two times the value after the ^ symbol (two times *g*) divided by the number before the ^ symbol: $\frac{2g}{h}$. Now, take that value, the value of *h*^*g*, and substitute it for *h*^*g* in (*h*^*g*)^*h*: $(\frac{2g}{h})$^*h*. Now, repeat the process. Two times the value after the ^ symbol (two times *h*) divided by the number before the symbol: $\frac{2h}{\frac{2g}{h}} = \frac{2h^2}{2g} = \frac{h^2}{g}$.

14. c. If four copy machines make 240 copies in three minutes, then five copy machines will make 240 copies in *x* minutes:

$(4)(240)(3) = (5)(240)(x)$

$2{,}880 = 1{,}200x$

$x = 2.4$

Five copy machines will make 240 copies in 2.4 minutes. Since there are 60 seconds in a minute, 0.4 of a minute is equal to (0.4)(60) = 24 seconds. The copies will be made in 2 minutes, 24 seconds.

15. d. 40% of $j = 0.4j$, 50% of $k = 0.5k$. If $0.4j = 0.5k$, then $j = \frac{0.5k}{0.4} = 1.25k$. *j* is equal to 125% of *k*, which means that *j* is 25% larger than *k*.

16. e. *FDCB* is a rectangle, which means that angle *D* is a right angle. Angle *ECD* is 60 degrees, which makes triangle *EDC* a 30-60-90 right triangle. The leg opposite the 60-degree angle is equal to $\sqrt{3}$ times the length of the leg opposite the 30-degree angle. Therefore, the length of side *DC* is equal to $\frac{6}{\sqrt{3}}$, or $2\sqrt{3}$. The hypotenuse of a 30-60-90 right triangle is equal to twice the length of the leg opposite the 30-degree angle, so the length of *EC* is $2(2\sqrt{3}) = 4\sqrt{3}$. Angle *DCB* is also a right angle, and triangle *ABC* is also a

30-60-60 right triangle. Since angle *ECD* is 60 degrees, angle *ECB* is equal to 90 – 60 = 30 degrees. Therefore, the length of *AC*, the hypotenuse of triangle *ABC*, is twice the length of *AB*: 2(10) = 20. The length of *AC* is 20 and the length of *EC* is $4\sqrt{3}$. Therefore, the length of *AE* is $20 - 4\sqrt{3}$.

CHAPTER

11 Practice Test 3

This practice test is a simulation of the three Math sections you will complete on the SAT. To receive the most benefit from this practice test, complete it as if it were the real SAT. So take this practice test under test-like conditions: Isolate yourself somewhere you will not be disturbed; use a stopwatch; follow the directions; and give yourself only the amount of time allotted for each section.

When you are finished, review the answers and explanations that immediately follow the test. Make note of the kinds of errors you made and review the appropriate skills and concepts before taking another practice test.

▶ Section 1

1. ⓐ ⓑ ⓒ ⓓ ⓔ
2. ⓐ ⓑ ⓒ ⓓ ⓔ
3. ⓐ ⓑ ⓒ ⓓ ⓔ
4. ⓐ ⓑ ⓒ ⓓ ⓔ
5. ⓐ ⓑ ⓒ ⓓ ⓔ
6. ⓐ ⓑ ⓒ ⓓ ⓔ
7. ⓐ ⓑ ⓒ ⓓ ⓔ
8. ⓐ ⓑ ⓒ ⓓ ⓔ
9. ⓐ ⓑ ⓒ ⓓ ⓔ
10. ⓐ ⓑ ⓒ ⓓ ⓔ
11. ⓐ ⓑ ⓒ ⓓ ⓔ
12. ⓐ ⓑ ⓒ ⓓ ⓔ
13. ⓐ ⓑ ⓒ ⓓ ⓔ
14. ⓐ ⓑ ⓒ ⓓ ⓔ
15. ⓐ ⓑ ⓒ ⓓ ⓔ
16. ⓐ ⓑ ⓒ ⓓ ⓔ
17. ⓐ ⓑ ⓒ ⓓ ⓔ
18. ⓐ ⓑ ⓒ ⓓ ⓔ
19. ⓐ ⓑ ⓒ ⓓ ⓔ
20. ⓐ ⓑ ⓒ ⓓ ⓔ

▶ Section 2

1. ⓐ ⓑ ⓒ ⓓ ⓔ
2. ⓐ ⓑ ⓒ ⓓ ⓔ
3. ⓐ ⓑ ⓒ ⓓ ⓔ
4. ⓐ ⓑ ⓒ ⓓ ⓔ
5. ⓐ ⓑ ⓒ ⓓ ⓔ
6. ⓐ ⓑ ⓒ ⓓ ⓔ
7. ⓐ ⓑ ⓒ ⓓ ⓔ
8. ⓐ ⓑ ⓒ ⓓ ⓔ

9.

10.

11.

12.

13.

14.

15.

16.

17.

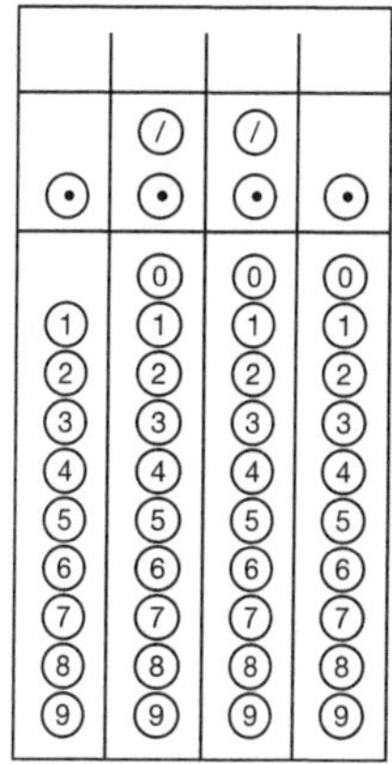

18.

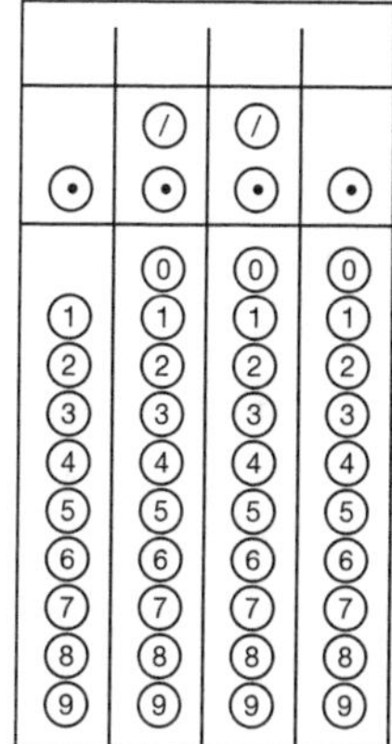

► Section 3

1. ⓐ ⓑ ⓒ ⓓ ⓔ
2. ⓐ ⓑ ⓒ ⓓ ⓔ
3. ⓐ ⓑ ⓒ ⓓ ⓔ
4. ⓐ ⓑ ⓒ ⓓ ⓔ
5. ⓐ ⓑ ⓒ ⓓ ⓔ
6. ⓐ ⓑ ⓒ ⓓ ⓔ
7. ⓐ ⓑ ⓒ ⓓ ⓔ
8. ⓐ ⓑ ⓒ ⓓ ⓔ
9. ⓐ ⓑ ⓒ ⓓ ⓔ
10. ⓐ ⓑ ⓒ ⓓ ⓔ
11. ⓐ ⓑ ⓒ ⓓ ⓔ
12. ⓐ ⓑ ⓒ ⓓ ⓔ
13. ⓐ ⓑ ⓒ ⓓ ⓔ
14. ⓐ ⓑ ⓒ ⓓ ⓔ
15. ⓐ ⓑ ⓒ ⓓ ⓔ
16. ⓐ ⓑ ⓒ ⓓ ⓔ

▶ Section 1

1. Which of the following could be equal to $\frac{x}{4x}$?

a. $-\frac{1}{4}$

b. $\frac{0}{4}$

c. 0.20

d. $\frac{4}{12}$

e. $\frac{5}{20}$

2. There are seven vocalists, four guitarists, four drummers, and two bassists in Glen Oak's music program, while there are five vocalists, eight guitarists, two drummers, and three bassists in Belmont's music program. If a band comprises one vocalist, one guitarist, one drummer, and one bassist, how many more bands can be formed in Belmont?

a. 4
b. 10
c. 16
d. 18
e. 26

3. Which of the following is the equation of a parabola whose vertex is at (5,–4)?

a. $y = (x - 5)^2 - 4$
b. $y = (x + 5)^2 - 4$
c. $y = (x - 5)^2 + 4$
d. $y = (x + 5)^2 + 4$
e. $y = x^2 - 29$

4. If $b^3 = -64$, then $b^2 - 3b - 4 =$

a. –6.
b. –4.
c. 0.
d. 24.
e. 28.

5.

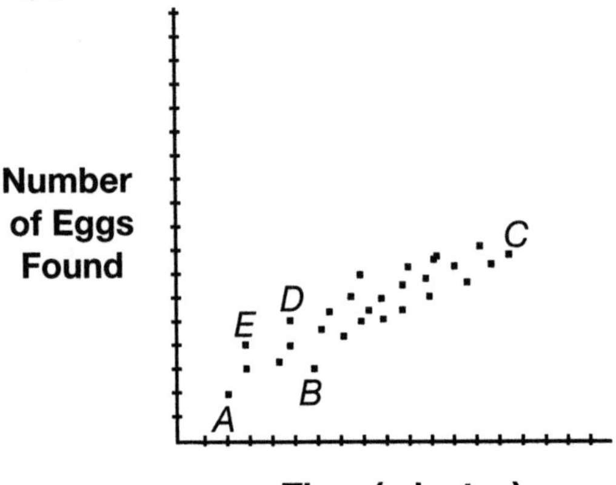

The scatter plot above shows how many eggs were found in a hunt over time. Which of the labeled points represents a number of eggs found that is greater than the number of minutes that has elapsed?

a. *A*
b. *B*
c. *C*
d. *D*
e. *E*

6. The point (6, –3) could be the midpoint of which of the following lines?

a. a line with endpoints at (0,–1) and (12,–2)
b. a line with endpoints at (2,–3) and (6,1)
c. a line with endpoints at (6,0) and (6,–6)
d. a line with endpoints at (–6,3) and (–6,–3)
e. a line with endpoints at (3,3) and (12,–6)

7. A sack contains red, blue, and yellow marbles. The ratio of red marbles to blue marbles to yellow marbles is 3:4:8. If there are 24 yellow marbles in the sack, how many total marbles are in the sack?

a. 45
b. 48
c. 72
d. 96
e. 144

8. What two values are not in the domain of $y = \frac{x^2 - 36}{x^2 - 9x - 36}$?

a. –3, 12
b. 3, –12
c. –6, 6
d. –6, 36
e. 9, 36

9. The diagonal of one face of a cube measures $4\sqrt{2}$ in. What is the volume of the cube?

a. $24\sqrt{2}$ in^3
b. 64 in^3
c. 96 in^3
d. $128\sqrt{2}$ in^3
e. 192 in^3

10. A line has a y-intercept of –6 and an x-intercept of 9. Which of the following is a point on the line?

a. (–6,–10)
b. (1,3)
c. (0,9)
d. (3,–8)
e. (6,13)

11. If $m < n < 0$, then all of the following are true EXCEPT

a. $-m < -n$.
b. $mn > 0$.
c. $|m| + n > 0$.
d. $|n| < |m|$.
e. $m - n < 0$.

12. The area of a circle is equal to four times its circumference. What is the circumference of the circle?

a. π units
b. 16π units
c. 48π units
d. 64π units
e. cannot be determined

13. If the statement "All students take the bus to school" is true, then which of the following must be true?

a. If Courtney does not take the bus to school, then she is not a student.
b. If Courtney takes the bus to school, then she is a student.
c. If Courtney is not a student, then she does not take the bus.
d. all of the above
e. none of the above

14.

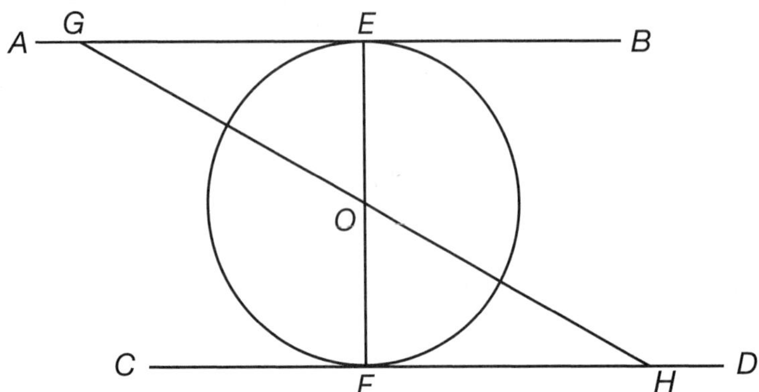

In the diagram above, line *AB* is parallel to line *CD*, both lines are tangents to circle *O* and the diameter of circle *O* is equal in measure to the length of line *OH*. If the diameter of circle *O* is 24 in, what is the measure of angle *BGH*?

a. 30 degrees
b. 45 degrees
c. 60 degrees
d. 75 degrees
e. cannot be determined

15.

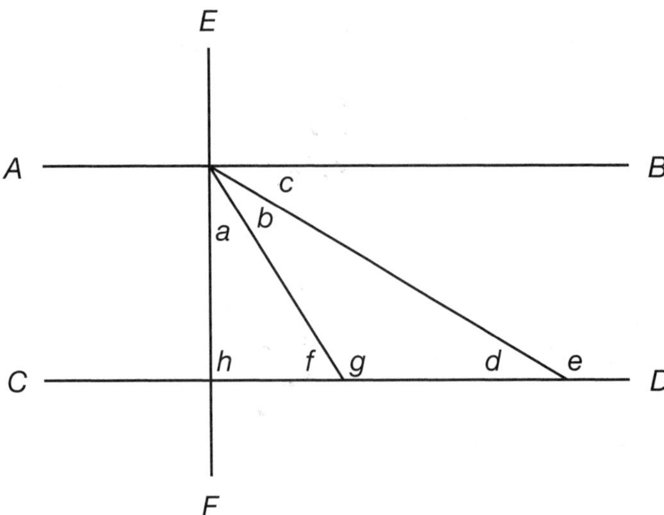

In the diagram above, if line *AB* is parallel to line *CD*, and line *EF* is perpendicular to lines *AB* and *CD*, all of the following are true EXCEPT

a. $e = a + b + 90$.
b. $a + h + f = b + g + d$.
c. $a + h = g$.
d. $a + b + d = 90$.
e. $c + b = g$.

16. If the lengths of the edges of a cube are decreased by 20%, the surface area of the cube will decrease by

a. 20%.
b. 36%.
c. 40%.
d. 51%.
e. 120%.

17. Simon plays a video game four times. His game scores are 18 points, 27 points, 12 points, and 15 points. How many points must Simon score in his fifth game in order for the mean, median, and mode of the five games to equal each other?

a. 12 points
b. 15 points
c. 18 points
d. 21 points
e. 27 points

18. If $g^{\frac{2}{5}} = 16$, then $g^{(-\frac{1}{5})} =$

a. $\frac{1}{4}$.
b. $\frac{1}{8}$.
c. $\frac{16}{5}$.
d. 4.
e. 8.

19.

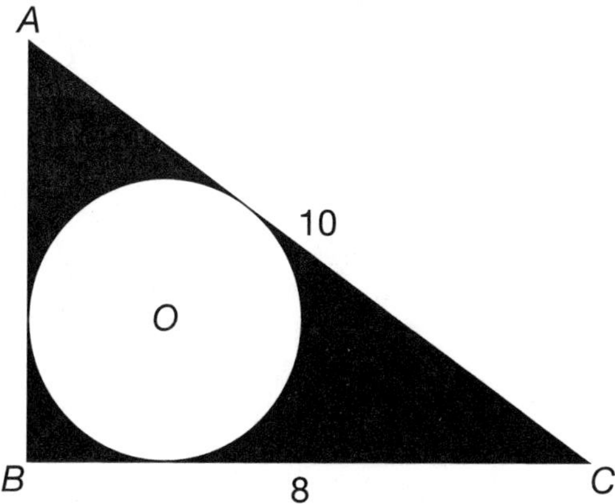

In the diagram above, triangle *ABC* is a right triangle and the diameter of circle *O* is $\frac{2}{3}$ the length of *AB*. Which of the following is equal to the shaded area?

a. 20π square units
b. $24 - 4\pi$ square units
c. $24 - 16\pi$ square units
d. $48 - 4\pi$ square units
e. $48 - 16\pi$ square units

20. In a restaurant, the ratio of four-person booths to two-person booths is 3:5. If 154 people can be seated in the restaurant, how many two-person booths are in the restaurant?

a. 14
b. 21
c. 35
d. 57
e. 70

▶ Section 2

1. If $y = -x^3 + 3x - 3$, what is the value of y when $x = -3$?

a. –35
b. –21
c. 15
d. 18
e. 33

2. What is the tenth term of the sequence: 5, 15, 45, 135 . . . ?

a. 5^{10}
b. $\frac{3^{10}}{5}$
c. $(5 \times 3)^9$
d. 5×3^9
e. 5×3^{10}

3. Wendy tutors math students after school every day for five days. Each day, she tutors twice as many students as she tutored the previous day. If she tutors t students the first day, what is the average (arithmetic mean) number of students she tutors each day over the course of the week?

a. t
b. $5t$
c. $6t$
d. $\frac{t^5}{5}$
e. $\frac{31t}{5}$

4. A pair of Jump sneakers costs $60 and a pair of Speed sneakers costs $45. For the two pairs of sneakers to be the same price

a. the price of a pair of Jump sneakers must decrease by 15%.
b. the price of a pair of Speed sneakers must increase by 15%.
c. the price of a pair of Jump sneakers must decrease by 25%.
d. the price of a pair of Speed sneakers must increase by 25%.
e. the price of a pair of Jump sneakers must decrease by 33%.

5.

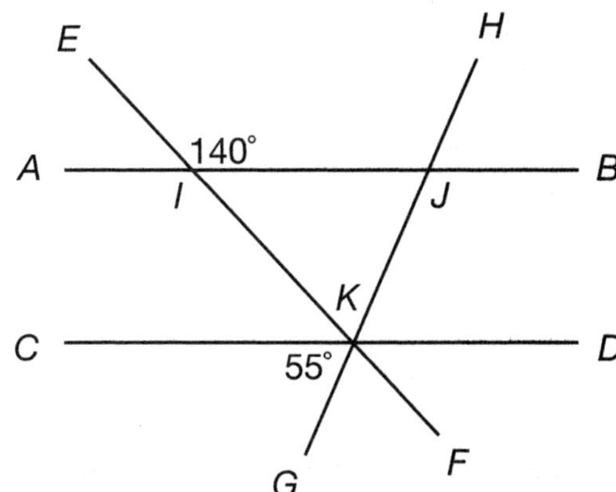

In the diagram above, line *AB* is parallel to line *CD*, angle *EIJ* measures 140 degrees and angle *CKG* measures 55 degrees. What is the measure of angle *IKJ*?

a. 40 degrees
b. 55 degrees
c. 85 degrees
d. 95 degrees
e. 135 degrees

6. A number cube is labeled with the numbers one through six, with one number on each side of the cube. What is the probability of rolling either a number that is even or a number that is a factor of 9?

a. $\frac{1}{3}$
b. $\frac{1}{2}$
c. $\frac{2}{3}$
d. $\frac{5}{6}$
e. 1

7. The area of one square face of a rectangular prism is 121 square units. If the volume of the prism is 968 cubic units, what is the surface area of the prism?

a. 352 square units
b. 512 square units
c. 528 square units
d. 594 square units
e. 1,452 square units

8.

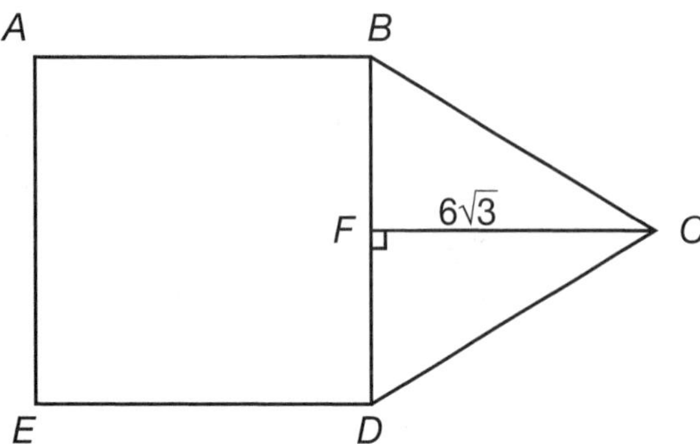

In the diagram above, *ABDE* is a square and *BCD* is an equilateral triangle. If $FC = 6\sqrt{3}$ cm, what is the perimeter of *ABCDE*?

a. $30\sqrt{3}$ cm
b. $36\sqrt{3}$ cm
c. 60 cm
d. $60\sqrt{3}$ cm
e. 84 cm

9. What is the value of $(3xy + x)\frac{x}{y}$ when $x = 2$ and $y = 5$?

10.

Ages of Spring Island Concert Attendees

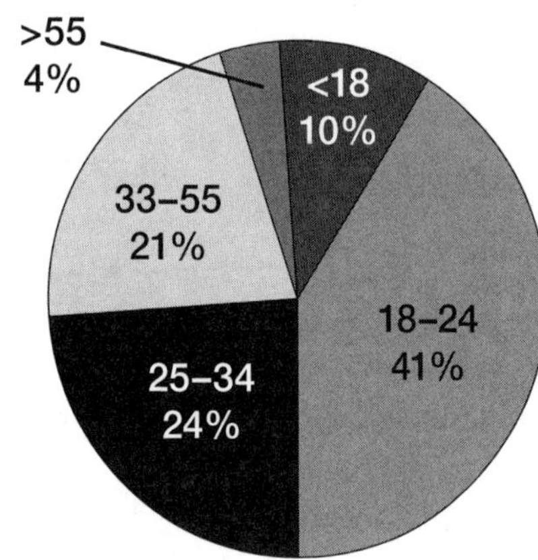

The diagram above shows the breakdown by age of the 1,560 people who attended the Spring Island Concert last weekend. How many people between the ages of 18 and 34 attended the concert?

11. Matt weighs $\frac{3}{5}$ of Paul's weight. If Matt were to gain 4.8 pounds, he would weigh $\frac{2}{3}$ of Paul's weight. What is Matt's weight in pounds?

12. If $-6b + 2a - 25 = 5$ and $\frac{a}{b} + 6 = 4$, what is the value of $(\frac{b}{a})^2$?

13. The function $j@k = (\frac{j}{k})^j$. If $j@k = -8$ when $j = -3$, what is the value of k?

14.

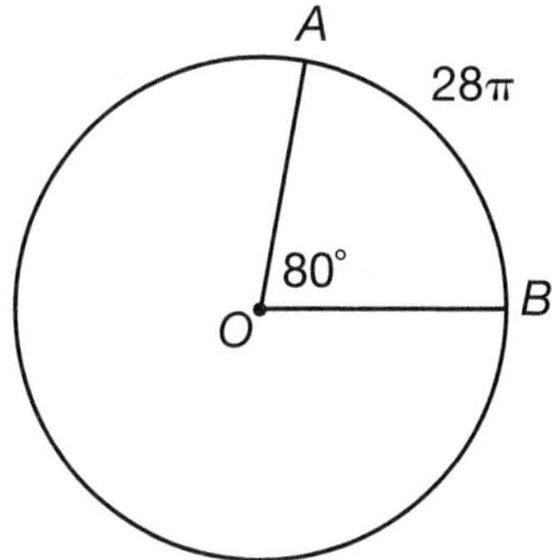

In the circle above, the measure of angle AOB is 80 degrees and the length of arc AB is 28π units. What is the radius of the circle?

15. What is the distance from the point where the line given by the equation $3y = 4x + 24$ crosses the x-axis to the point where the line crosses the y-axis?

16. For any whole number $x > 0$, how many elements are in the set that contains only the numbers that are multiples AND factors of x?

17. A bus holds 68 people. If there must be one adult for every four children on the bus, how many children can fit on the bus?

18. In Marie's fish tank, the ratio of guppies to platies is 4:5. She adds nine guppies to her fish tank and the ratio of guppies to platies becomes 5:4. How many guppies are in the fish tank now?

▶ Section 3

1. The line $y = -2x + 8$ is
a. parallel to the line $y = \frac{1}{2}x + 8$.
b. parallel to the line $\frac{1}{2}y = -x + 3$.
c. perpendicular to the line $2y = -\frac{1}{2}x + 8$.
d. perpendicular to the line $\frac{1}{2}y = -2x - 8$.
e. perpendicular to the line $y = 2x - 8$.

2. It takes six people eight hours to stuff 10,000 envelopes. How many people would be required to do the job in three hours?
a. 4
b. 12
c. 16
d. 18
e. 24

3.

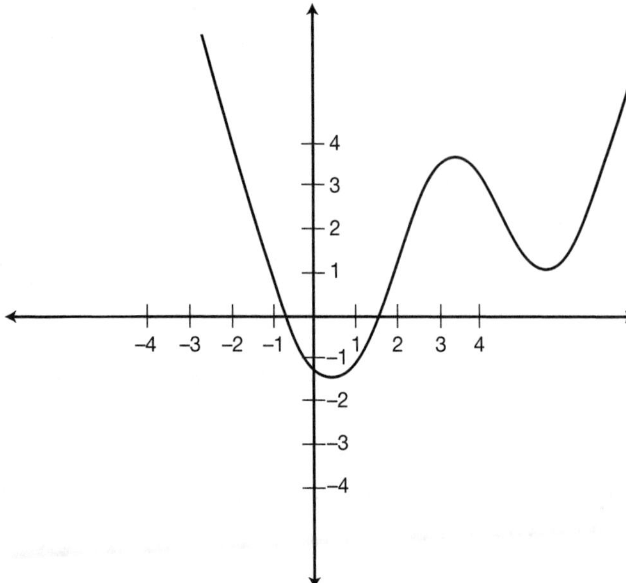

In the diagram above of $f(x)$, for how many values does $f(x) = -1$?

a. 0
b. 1
c. 2
d. 3
e. 4

4. The equation $\frac{x^2}{4} - 3x = -8$ when $x =$

a. –8 or 8.
b. –4 or 4.
c. –4 or –8.
d. 4 or –8.
e. 4 or 8.

5. The expression $\frac{x^2 - 16}{x^3 + x^2 - 20x}$ can be reduced to

a. $\frac{4}{x+5}$.
b. $\frac{x+4}{x}$.
c. $\frac{x+4}{x+5}$.
d. $\frac{x+4}{x^2+5x}$.
e. $-\frac{16}{x3-20x}$.

6.

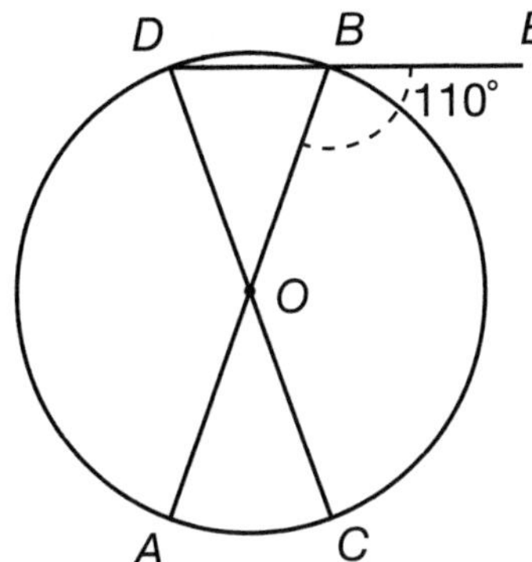

In the diagram above, if angle *OBE* measures 110 degrees, what is the measure of arc *AC*?

a. 20 degrees
b. 40 degrees
c. 55 degrees
d. 80 degrees
e. cannot be determined

7. The volume of a cylinder is 486π cubic units. If the height of the cylinder is six units, what is the total area of the bases of the cylinder?

a. 9π square units
b. 18π square units
c. 27π square units
d. 81π square units
e. 162π square units

8. If $a\sqrt{20} = \frac{2\sqrt{180}}{a}$, then $a =$

a. $2\sqrt{3}$.
b. $\sqrt{5}$.
c. 5.
d. $\sqrt{6}$.
e. 6.

9.

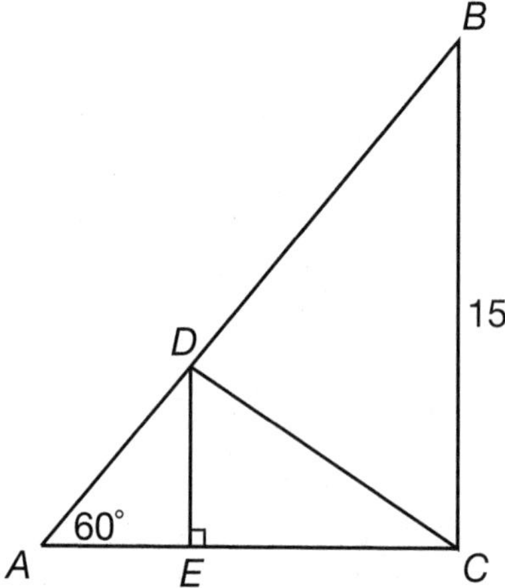

In the diagram above, ABC and DEC are right triangles, the length of side BC is 15 units, and the measure of angle A is 60 degrees. If angle A is congruent to angle EDC, what is the length of side DC?

a. $\sqrt{15}$ units

b. $\frac{15}{2}$ units

c. $\frac{15}{2}\sqrt{3}$ units

d. 9 units

e. $15\sqrt{3}$ units

10. If q is decreased by p percent, then the value of q is now

a. $q - p$.

b. $q - \frac{p}{100}$.

c. $-\frac{pq}{100}$.

d. $q - \frac{pq}{100}$.

e. $pq - \frac{pq}{100}$.

11. The product of $(\frac{a}{b})^2(\frac{b}{a})^{-2}(\frac{1}{a})^{-1} =$

a. a.

b. $\frac{1}{a}$.

c. $\frac{a^3}{b^4}$.

d. $\frac{a^4}{b^4}$.

e. $\frac{a^5}{b^4}$.

12. Gil drives five times farther in 40 minutes than Warrick drives in 30 minutes. If Gil drives 45 miles per hour, how fast does Warrick drive?

a. 6 mph
b. 9 mph
c. 12 mph
d. 15 mph
e. 30 mph

13. A bank contains one penny, two quarters, four nickels, and three dimes. What is the probability of selecting a coin that is worth more than five cents but less than 30 cents?

a. $\frac{1}{5}$
b. $\frac{1}{4}$
c. $\frac{1}{2}$
d. $\frac{7}{10}$
e. $\frac{9}{10}$

14.

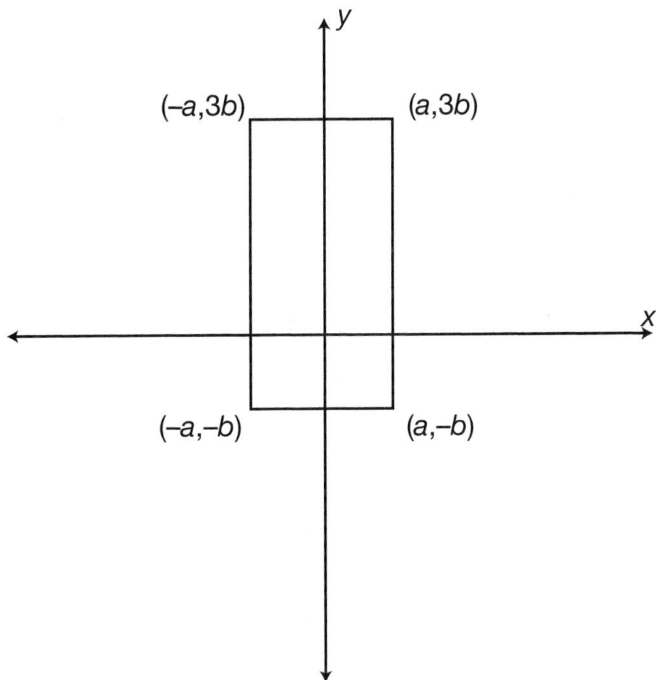

In the diagram above, what is the area of the rectangle?

a. $6ab$ square units
b. $8ab$ square units
c. $9b^2$ square units
d. $12ab$ square units
e. $16b$ square units

15. If set M contains only the positive factors of 8 and set N contains only the positive factors of 16, then the union of sets M and N

a. contains exactly the same elements that are in set N.
b. contains only the elements that are in both sets M and N.
c. contains nine elements.
d. contains four elements.
e. contains only even elements.

16.

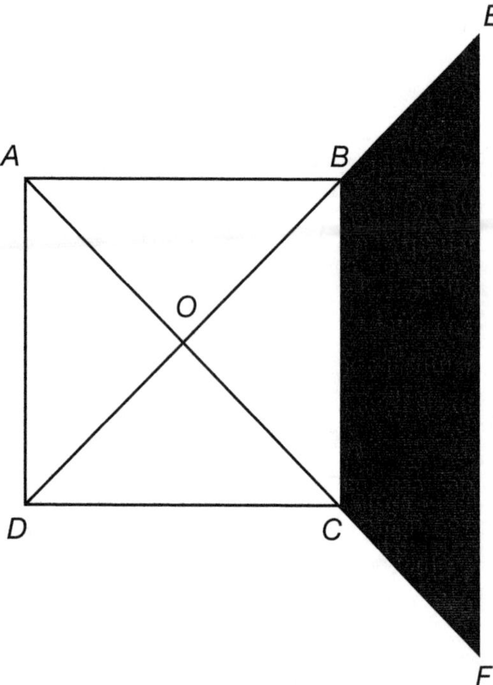

In the diagram above, $ABCD$ is a square with an area of 100 cm^2 and lines BD and AC are the diagonals of $ABCD$. If line EF is parallel to line BC and the length of line $CF = 3\sqrt{2}$ cm, which of the following is equal to the shaded area?

a. 25 cm^2
b. 39 cm^2
c. 64 cm^2
d. 78 cm^2
e. 89 cm^2

▶ Answer Key

Section 1 Answers

1. e. Divide the numerator and denominator of $\frac{x}{4x}$ by x, leaving $\frac{1}{4}$. Divide the numerator and denominator of $\frac{5}{20}$ by 5. This fraction is also equal to $\frac{1}{4}$.

2. c. Multiply the numbers of vocalists, guitarists, drummers, and bassists in each town to find the number of bands that can be formed in each town. There are $(7)(4)(4)(2) = 224$ bands that can be formed in Glen Oak. There are $(5)(8)(2)(3) = 240$ bands that can be formed in Belmont; $240 - 224 = 16$ more bands that can be formed in Belmont.

3. a. The equation of a parabola with its turning point five units to the right of the y-axis is written as $y = (x-5)^2$. The equation of a parabola with its turning point four units below the x-axis is written as $y = x^2 - 4$. Therefore, the equation of a parabola with its vertex at (5,–4) is $y = (x-5)^2 - 4$.

4. d. If $b^3 = -64$, then, taking the cube root of both sides, $b = -4$. Substitute –4 for b in the second equation: $b^2 - 3b - 4 = (-4)^2 - 3(-4) - 4 = 16 + 12 - 4 = 24$.

5. e. The point that represents a number of eggs found that is greater than the number of minutes that has elapsed is the point that has a y value that is greater than its x value. Only point E lies farther from the horizontal axis than it lies from the vertical axis. At point E, more eggs have been found than the number of minutes that has elapsed.

6. c. The midpoint of a line is equal to the average of the x-coordinates and the average of the y-coordinates of the endpoints of the line. The midpoint of the line with endpoints at (6,0) and (6,–6) is $(\frac{6+6}{2},\frac{0+-6}{2}) = (\frac{12}{2},-\frac{6}{2}) = (6,-3)$.

7. a. The number of yellow marbles, 24, is $\frac{24}{8} = 3$ times larger than the number of marbles given in the ratio. Multiply each number in the ratio by 3 to find the number of each color of marbles. There are $3(3) = 9$ red marbles and $4(3) = 12$ blue marbles. The total number of marbles in the sack is $24 + 9 + 12 = 45$.

8. a. The equation $y = \frac{x^2 - 36}{x^2 - 9x - 36}$ is undefined when its denominator, $x^2 - 9x - 36$, evaluates to zero. The x values that make the denominator evaluate to zero are not in the domain of the equation. Factor $x^2 - 9x - 36$ and set the factors equal to zero: $x^2 - 9x - 36 = (x - 12)(x + 3)$; $x - 12 = 0$, $x = 12$; $x + 3 = 0$, $x = -3$.

9. b. Every face of a cube is a square. The diagonal of a square is equal to $s\sqrt{2}$, where s is the length of a side of the square. If $s\sqrt{2} = 4\sqrt{2}$, then one side, or edge, of the cube is equal to 4 in. The volume of a cube is equal to e^3, where e is the length of an edge of the cube. The volume of the cube is equal to $(4 \text{ in})^3 = 64 \text{ in}^3$.

10. a. A line with a y-intercept of –6 passes through the point (0,–6) and a line with an x-intercept of 9 passes through the point (9,0). The slope of a line is equal to the change in y values between two points on the line divided by the change in the x values of those points. The slope of this line is equal to $\frac{0-(-6)}{9-0} = \frac{6}{9} = \frac{2}{3}$. The equation of the line that has a slope of $\frac{2}{3}$ and a y-intercept of –6 is $y = \frac{2}{3}x - 6$. When $x = -6$, y is equal to $\frac{2}{3}(-6) - 6 = -4 - 6 = -10$; therefore, the point (–6,–10) is on the line $y = \frac{2}{3}x - 6$.

11. a. If $m < n < 0$, then m and n are both negative numbers, and m is more negative than n. Therefore, $-m$ will be more positive (greater) than $-n$, so the statement $-m < -n$ cannot be true.

12. b. If r is the radius of this circle, then the area of this circle, πr^2, is equal to four times its circumference, $2\pi r$: $\pi r^2 = 4(2\pi r)$, $\pi r^2 = 8\pi r$, $r^2 = 8r$, $r = 8$ units. If the radius of the circle is eight units, then its circumference is equal to $2\pi(8) = 16\pi$ units.

13. a. Since all students take the bus to school, anyone who does not take the bus cannot be a student. If Courtney does not take the bus to school, then she cannot be a student. However, it is not necessarily true that everyone who takes the bus to school is a student, nor is it necessarily true that everyone who is not a student does not take the bus. The statement "All students take the bus to school" does not, for instance, preclude the statement "Some teachers take the bus to school" from being true.

14. a. Lines OF and OE are radii of circle O and since a tangent and a radius form a right angle, triangles OFH and OGE are right triangles. If the length of the diameter of the circle is 24 in, then the length of the radius is 12 in. The sine of angle OHF is equal to $\frac{12}{24}$, or $\frac{1}{2}$. The measure of an angle with a sine of $\frac{1}{2}$ is 30 degrees. Therefore, angle OHF measures 30 degrees. Since angles BGH and OHF are alternating angles, they are equal in measure. Therefore, angle BGH also measures 30 degrees.

15. e. Since AB and CD are parallel lines cut by a transversal, angle f is equal to the sum of angles c and b. However, angle f and angle g are not equal—they are supplementary. Therefore, the sum of angles c and b is also supplementary—and not equal—to g.

16. b. The surface area of a cube is equal to $6e^2$, where e is the length of an edge of a cube. The surface area of a cube with an edge equal to one unit is 6 cubic units. If the lengths of the edges are decreased by 20%, then the surface area becomes $6(\frac{4}{5})^2 = \frac{96}{25}$ cubic units, a decrease of $\frac{6-\frac{96}{25}}{6} = \frac{\frac{54}{25}}{6} = \frac{9}{25} = \frac{36}{100} = 36\%$.

17. c. For the median and mode to equal each other, the fifth score must be the same as one of the first four, and, it must fall in the middle position when the five scores are ordered. Therefore, Simon must have scored either 15 or 18 points in his fifth game. If he scored 15 points, then his mean score would have been greater than 15: 17.4. Simon scored 18 points in his fifth game, making the mean, median, and mode for the five games equal to 18.

18. a. To go from $g(\frac{2}{5})$ to $g(-\frac{1}{5})$, you would multiply the exponent of $g(\frac{2}{5})$ by $(-\frac{1}{2})$. Therefore, to go from 16 (the value of $g(\frac{2}{5})$) to the value of $g(-\frac{1}{5})$, multiply the exponent of 16 by $(-\frac{1}{2})$. The exponent of 16 is one, so the value of $g(-\frac{1}{5}) = 16$ to the $(-\frac{1}{2})$ power, which is $\frac{1}{4}$.

19. b. Since ABC is a right triangle, the sum of the squares of its legs is equal to the square of the hypotenuse: $(AB)^2 + 8^2 = 10^2$, $(AB)^2 + 64 = 100$, $(AB)^2 = 36$, $AB = 6$ units. The diameter of circle O is $\frac{2}{3}$ of AB, or $\frac{2}{3}(6) = 4$ units. The area of a triangle is equal to $\frac{1}{2}bh$, where b is the base of the triangle and h is the height of the triangle. The area of $ABC = \frac{1}{2}(6)(8) = 24$ square units. The area of a circle is equal to πr^2, where r is the radius of the circle. The radius of a circle is equal to half the diameter of the circle, so the radius of O is $\frac{1}{2}(4) = 2$ units. The area of circle $O = \pi(2)^2 = 4\pi$. The shaded area is equal to the area of the triangle minus the area of the circle: $24 - 4\pi$ square units.

20. c. Let $3x$ equal the number of four-person booths and let $5x$ equal the number of two-person booths. Each four-person booth holds four people and each two-person booth holds two people. Therefore, $(3x)(4) + (5x)(2) = 154$, $12x + 10x = 154$, $22x = 154$, $x = 7$. There are $(7)(3) = 21$ four-person booths and $(7)(5) = 35$ two-person booths.

Section 2 Answers

1. c. Substitute –3 for x and solve for y:

$y = -(-3)^3 + 3(-3) - 3$

$y = -(-27) - 9 - 3$

$y = 27 - 12$

$y = 15$

2. d. The first term in the sequence is equal to 5×3^0, the second term is equal to 5×3^1, and so on. Each term in the pattern is equal to $5 \times 3^{(n-1)}$, where n is the position of the term in the pattern. The tenth term in the pattern is equal to $5 \times 3^{(10-1)}$, or 5×3^9.

3. e. If Wendy tutors t students the first day, then she tutors $2t$ students the second day, $4t$ students the third day, $8t$ students the fourth day, and $16t$ students the fifth day. The average number of students tutored each day over the course of the week is equal to the sum of the tutored students divided by the number of days: $\frac{t + 2t + 4t + 8t + 16t}{5} = \frac{31t}{5}$.

4. c. Jump sneakers cost \$60 – \$45 = \$15 more, or $\frac{15}{45}$ = 33% more than Speed sneakers. Speed sneakers cost \$15 less, or $\frac{15}{60}$ = 25% less than Jump sneakers. For the two pairs of sneakers to be the same price, either the price of Speed sneakers must increase by 33% or the price of Jump sneakers must decrease by 25%.

5. c. Since *AB* and *CD* are parallel lines cut by transversals *EF* and *GH* respectively, angles *CKG* and *IJK* are alternating angles. Alternating angles are equal in measure, so angle *IJK* = 55 degrees. Angles *EIJ* and *JIK* form a line. They are supplementary and their measures sum to 180 degrees. Angle *JIK* = 180 – 140 = 40 degrees. Angles *JIK*, *IJK*, and *IKJ* comprise a triangle. There are 180 degrees in a triangle; therefore, the measure of angle *IKJ* = 180 – (55 + 40) = 85 degrees.

6. d. There are three numbers on the cube that are even (2, 4, 6), so the probability of rolling an even number is $\frac{1}{2}$. There are two numbers on the cube that are factors of 9 (1, 3), so the probability of rolling a factor of 9 is $\frac{2}{6}$ or $\frac{1}{3}$. No numbers are members of both sets, so to find the probability of rolling either a number that is even or a number that is a factor of 9, add the probability of each event: $\frac{1}{2} + \frac{1}{3} = \frac{3}{6} + \frac{2}{6} = \frac{5}{6}$.

7. d. The area of a square is equal to the length of a side, or edge, of the square times itself. If the area of a square face is 121 square units, then the lengths of two edges of the prism are 11 units. The volume of the prism is 968 cubic units. The volume of prism is equal to *lwh*, where *l* is the length of the prism, *w* is the width of the prism, and *h* is the height of the prism. The length and width of the prism are both 11 units. The height is equal to: 968 = (11)(11)h, 968 = 121h, h = 8. The prism has two square faces and four rectangular faces. The area of one square face is 121 square units. The area of one rectangular face is (8)(11) = 88 square units. Therefore, the total surface area of the prism is equal to: 2(121) + 4(88) = 242 + 352 = 594 square units.

8. c. Since *BCD* is an equilateral triangle, angles *CBD*, *BDC*, and *BCD* all measure 60 degrees. *FCD* and *BCF* are both 30-60-90 right triangles that are congruent to each other. The side opposite the 60-degree angle of triangle *BCF*, side *FC*, is equal to $\sqrt{3}$ times the length of the side opposite the 30-degree angle, side *BF*. Therefore, *BF* is equal to = 6 cm. The hypotenuse, *BC*, is equal to twice the length of side *BF*. The length of *BC* is 2(6) = 12 cm. Since *BC* = 12 cm, *CD* and *BD* are also 12 cm. *BD* is one side of square *ABDE*; therefore, each side of *ABDE* is equal to 12 cm. The perimeter of *ABCDE* = 12 cm + 12 cm + 12 cm + 12 cm + 12 cm = 60 cm.

9. 4 Substitute 2 for x and 5 for y: $(3xy + x)^{\frac{x}{y}} = ((3)(2)(5) + 2)^{\frac{2}{5}} = (30 + 2)^{\frac{2}{5}} = 32^{\frac{2}{5}} = (\sqrt[5]{32})^2 = 2^2 = 4$. Or, 3(2)(5) = 30, 30 + 2 = 32, the 5th root of 32 is 2, 2 raised to the 2nd power is 4.

10. 1,014 Of the concert attendees, 41% were between the ages of 18–24 and 24% were between the ages of 25–34. Therefore, 41 + 24 = 65% of the attendees, or (1,560)(0.65) = 1,014 people between the ages of 18 and 34 attended the concert.

11. 43.2 Matt's weight, m, is equal to $\frac{3}{5}$ of Paul's weight, p: $m = \frac{3}{5}p$. If 4.8 is added to m, the sum is equal to $\frac{2}{3}$ of p: $m + 4.8 = \frac{2}{3}p$. Substitute the value of m in terms of p into the second equation: $\frac{3}{5}p + 4.8 = \frac{2}{3}p$, $\frac{1}{15}p = 4.8$, $p = 72$. Paul weighs 72 pounds, and Matt weighs $\frac{3}{5}(72) = 43.2$ pounds.

12. $\frac{1}{4}$ Solve $-6b + 2a - 25 = 5$ for a in terms of b: $-6b + 2a - 25 = 5$, $-3b + a = 15$, $a = 15 + 3b$. Substitute a in terms of b into the second equation: $\frac{15 + 3b}{b} + 6 = 4$, $\frac{15}{b} + 3 + 6 = 4$, $\frac{15}{b} = -5$, $b = -3$. Substitute b into the first equation to find the value of a: $-6b + 2a - 25 = 5$, $-6(-3) + 2a - 25 = 5$, $18 + 2a = 30$, $2a = 12$, $a = 6$. Finally, $(\frac{b}{a})^2 = (\frac{-3}{6})^2 = (-\frac{1}{2})^2 = \frac{1}{4}$.

13. 6 If $j@k = -8$ when $j = -3$, then:

$-8 = (\frac{-3}{k})^{-3}$

$-8 = (\frac{k}{-3})^3$

$-8 = -\frac{k^3}{27}$

$216 = k^3$

$k = 6$

14. 63 The size of an intercepted arc is equal to the measure of the intercepting angle divided by 360, multiplied by the circumference of the circle ($2\pi r$, where r is the radius of the circle): $28\pi = (\frac{80}{360})(2\pi r)$, $28 = (\frac{4}{9})r$, $r = 63$ units.

15. 10 Write the equation in slope-intercept form ($y = mx + b$): $3y = 4x + 24$, $y = \frac{4}{3}x + 8$. The line crosses the y-axis at its y-intercept, (0,8). The line crosses the x-axis when $y = 0$: $\frac{4}{3}x + 8 = 0$, $\frac{4}{3}x = -8$, $x = -6$. Use the distance formula to find the distance from (0,8) to (–6,0):

Distance = $\sqrt{(x_2 - x_1)^2 + (y_2 - y_1)^2}$

Distance = $\sqrt{((-6) - 0)^2 + (0 - 8)^2}$

Distance = $\sqrt{6^2 + (-8)^2}$

Distance = $\sqrt{36 + 64}$

Distance = $\sqrt{100}$

Distance = 10 units.

16. 1 The largest factor of a positive, whole number is itself, and the smallest multiple of a positive, whole number is itself. Therefore, the set of only the factors and multiples of a positive, whole number contains one element—the number itself.

17. 52 There is one adult for every four children on the bus. Divide the size of the bus, 68, by 5: $\frac{68}{5} = 13.6$. There can be no more than 13 groups of one adult, four children. Therefore, there can be no more than (13 groups)(4 children in a group) = 52 children on the bus.

18. 25 If the original ratio of guppies, g, to platies, p, is 4:5, then $g = \frac{4}{5}p$. If nine guppies are added, then the new number of guppies, $g + 9$, is equal to $\frac{5}{4}p$: $g + 9 = \frac{5}{4}p$. Substitute the value of g in terms of p from the first equation: $\frac{4}{5}p + 9 = \frac{5}{4}p$, $9 = \frac{9}{20}p$, $p = 20$. There are 20 platies in the fish tank and there are now $20(\frac{5}{4}) = 25$ guppies in the fish tank. $\frac{6\sqrt{3}}{\sqrt{3}}$

Section 3 Answers

1. b. Parallel lines have the same slope. When an equation is written in the form $y = mx + b$, the value of m (the coefficient of x) is the slope. The line $y = -2x + 8$ has a slope of –2. The line $\frac{1}{2}y = -x + 3$ is equal to $y = -2x + 6$. This line has the same slope as the line $y = -2x + 8$; therefore, these lines are parallel.

2. c. Six people working eight hours produce (6)(8) = 48 work-hours. The number of people required to produce 48 work-hours in three hours is $\frac{48}{3} = 16$.

3. c. The function $f(x)$ is equal to –1 every time the graph of $f(x)$ crosses the line $y = -1$. The graph of $f(x)$ crosses $y = -1$ twice; therefore, there are two values for which $f(x) = -1$.

4. e. Write the equation in quadratic form and find its roots:

$\frac{x^2}{4} - 3x = -8$

$x^2 - 12x = -32$

$x^2 - 12x + 32 = 0$

$(x - 8)(x - 4) = 0$

$x - 8 = 0, x = 8$

$x - 4 = 0, x = 4$

$\frac{x^2}{4} - 3x = -8$ when x is either 4 or 8.

5. d. Factor the numerator and denominator; $x^2 - 16 = (x + 4)(x - 4)$ and $x^3 + x^2 - 20x = x(x + 5)(x - 4)$. Cancel the $(x - 4)$ terms that appear in the numerator and denominator. The fraction becomes $\frac{x+4}{x(x+5)}$, or $\frac{x+4}{x^2+5x}$.

6. b. Angles *OBE* and *DBO* form a line. Since there are 180 degrees in a line, the measure of angle *DBO* is 180 – 110 = 70 degrees. *OB* and *DO* are radii, which makes triangle *DBO* isosceles, and angles *ODB* and *DBO* congruent. Since *DBO* is 70 degrees, *ODB* is also 70 degrees, and *DOB* is 180 – (70 + 70) = 180 – 140 = 40 degrees. Angles *DOB* and *AOC* are vertical angles, so the measure of angle *AOC* is also 40 degrees. Angle *AOC* is a central angle, so its intercepted arc, *AC*, also measures 40 degrees.

7. e. The volume of a cylinder is equal to $\pi r^2 h$, where r is the radius of the cylinder and h is the height of the cylinder. If the height of a cylinder with a volume of 486π cubic units is six units, then the radius is equal to:

$486\pi = \pi r^2(6)$

$486 = 6r^2$

$81 = r^2$

$r = 9$

A cylinder has two circular bases. The area of a circle is equal to πr^2, so the total area of the bases of the cylinder is equal to $2\pi r^2$, or $2\pi(9)^2 = 2(81)\pi = 162\pi$ square units.

8. d. Cross multiply:

$a\sqrt{20} = \frac{2\sqrt{180}}{a}$

$a^2\sqrt{20} = 2\sqrt{180}$

$a^2\sqrt{4}\sqrt{5} = 2\sqrt{36}\sqrt{5}$

$2a^2\sqrt{5} = 12\sqrt{5}$

$a^2 = 6$

$a = \sqrt{6}$

9. b. Since triangle *DEC* is a right triangle, triangle *AED* is also a right triangle, with a right angle at *AED*. There are 180 degrees in a triangle, so the measure of angle *ADE* is 180 – (60 + 90) = 30 degrees. Angle *A* and angle *EDC* are congruent, so angle *EDC* is also 60 degrees. Since there are 180 degrees in a line, angle *BDC* must be 90 degrees, making triangle *BDC* a right triangle. Triangle *ABC* is a right triangle with angle *A* measuring 60 degrees, which means that angle *B* must be 30 degrees, and *BDC* must be a 30-60-90 right triangle. The leg opposite the 30-degree angle in a 30-60-90 right triangle is half the length of the hypotenuse. Therefore, the length of *DC* is $\frac{15}{2}$ units.

10. d. p percent of q is equal to $q(\frac{p}{100})$, or $\frac{pq}{100}$. If q is decreased by this amount, then the value of q is $\frac{pq}{100}$ less than q, or $q - \frac{pq}{100}$.

11. e. A fraction with a negative exponent can be rewritten as a fraction with a positive exponent by switching the numerator with the denominator.

$(\frac{a}{b})^2(\frac{b}{a})^{-2}(\frac{1}{a})^{-1} = (\frac{a}{b})^2(\frac{a}{b})^2(\frac{a}{1})^1 = (\frac{a^2}{b^2})(\frac{a^2}{b^2})(a) = \frac{a^5}{b^4}$.

12. c. If d is the distance Warrick drives and s is the speed Warrick drives, then $30s = d$. Gil drives five times farther, $5d$, in 40 minutes, traveling 45 miles per hour: $5d = (40)(45)$. Substitute the value of d in terms of s into the second equation and solve for s, Warrick's speed: $5(30s) = (40)(45)$, $150s = 1{,}800$, $s = 12$. Warrick drives 12 mph.

13. c. There are ten coins in the bank (1 penny + 2 quarters + 4 nickels + 3 dimes). The two quarters and three dimes are each worth more than five cents but less than 30 cents, so the probability of selecting one of these coins is $\frac{5}{10}$ or $\frac{1}{2}$.

14. b. The y-axis divides the rectangle in half. Half of the width of the rectangle is a units to the left of the y-axis and the other half is a units to the right of the y-axis. Therefore, the width of the rectangle is $2a$ units. The length of the rectangle stretches from $3b$ units above the x-axis to b units below the x-axis. Therefore, the length of the rectangle is $4b$ units. The area of a rectangle is equal to lw, where l is the length of the rectangle and w is the width of the rectangle. The area of this rectangle is equal to $(2a)(4b) = 8ab$ square units.

15. a. Set M contains the positive factors of 8: 1, 2, 4, and 8. Set N contains the positive factors of 16: 1, 2, 4, 8, and 16. The union of these sets is equal to all of the elements that are in either set. Since every element in set M is in set N, the union of N and M is the same as set N: {1, 2, 4, 8, 16}.

16. b. The area of a square is equal to s^2, where s is the length of one side of the square. A square with an area of 100 cm^2 has sides that are each equal to $\sqrt{100} = 10$ cm. The diagonal of a square is equal to $\sqrt{2}$ times the length of a side of the square. Therefore, the lengths of diagonals AC and BD are $10\sqrt{2}$ cm. Diagonals of a square bisect each other at right angles, so the lengths of segments OB and OC are each $5\sqrt{2}$ cm. Since lines BC and EF are parallel and lines OC and OB are congruent, lines BE and CF are also congruent. The length of line OF is equal to the length of line OC plus the length of line CF: $5\sqrt{2} + 3\sqrt{2} = 8\sqrt{2}$ cm. In the same way, $OE = OB + BE = 5\sqrt{2} + 3\sqrt{2} = 8\sqrt{2}$ cm. The area of a triangle is equal to $\frac{1}{2}bh$, where b is the base of the triangle and h is the height of the triangle. EOF is a right triangle, and its area is equal to $\frac{1}{2}(8\sqrt{2})(8\sqrt{2}) = \frac{1}{2}(64)(2) = 64$ cm^2. The size of the shaded area is equal to the area of EOF minus one-fourth of the area of $ABCD$: $64 - \frac{1}{4}(100) = 64 - 25 = 39$ cm^2.

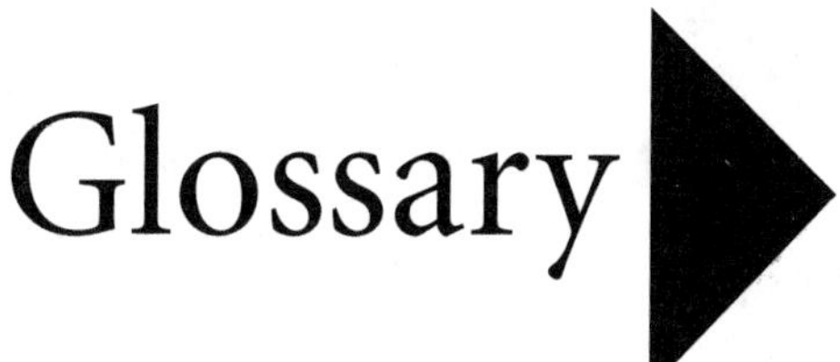

Glossary

absolute value the distance a number or expression is from zero on a number line

acute angle an angle that measures less than 90°

acute triangle a triangle with every angle that measures less than 90°

adjacent angles two angles that have the same vertex, share one side, and do not overlap

angle two rays connected by a vertex

arc a curved section of a circle

area the number of square units inside a shape

associative property of addition when adding three or more addends, the grouping of the addends does not affect the sum.

associative property of multiplication when multiplying three or more factors, the grouping of the factors does not affect the product.

average the quantity found by adding all the numbers in a set and dividing the sum by the number of addends; also known as the *mean*

base a number used as a repeated factor in an exponential expression. In 5^7, 5 is the base.

binomial a polynomial with two unlike terms, such as $2x + 4y$

bisect divide into two equal parts

central angle an angle formed by an arc in a circle

chord a line segment that goes through a circle, with its endpoints on the circle

circumference the distance around a circle

coefficient a number placed next to a variable

combination the arrangement of a group of items in which the order doesn't matter

common factors the factors shared by two or more numbers

common multiples multiples shared by two or more numbers

commutative property of addition when using addition, the order of the addends does not affect the sum.

commutative property of multiplication when using multiplication, the order of the factors does not affect the product.
complementary angles two angles whose sum is 90°
composite number a number that has more than two factors
congruent identical in shape and size; the geometric symbol for *congruent to* is ≅.
coordinate plane a grid divided into four quadrants by both a horizontal *x*-axis and a vertical *y*-axis
coordinate points points located on a coordinate plane
cross product a product of the numerator of one fraction and the denominator of a second fraction
denominator the bottom number in a fraction. 7 is the denominator of $\frac{3}{7}$.
diagonal a line segment between two non-adjacent vertices of a polygon
diameter a chord that passes through the center of a circle—the longest line you can draw in a circle. The term is used not only for this line segment, but also for its length.
difference the result of subtracting one number from another
distributive property when multiplying a sum (or a difference) by a third number, you can multiply each of the first two numbers by the third number and then add (or subtract) the products.
dividend a number that is divided by another number
divisor a number that is divided into another number
domain all the *x* values of a function
equation a mathematical statement that contains an equal sign
equiangular polygon a polygon with all angles of equal measure
equidistant the same distance
equilateral triangle a triangle with three equal sides and three equal angles
even number a number that can be divided evenly by the number 2 (resulting in a whole number)
exponent a number that tells you how many times a number, the base, is a factor in the product. In 5^7, 7 is the exponent.
exterior angle an angle on the outer sides of two lines cut by a transversal; or, an angle outside a triangle
factor a number that is multiplied to find a product
function a relationship in which one value depends upon another value
geometric sequence a sequence that has a constant ratio between terms
greatest common factor the largest of all the common factors of two or more numbers
hypotenuse the longest leg of a right triangle. The hypotenuse is always opposite the right angle in a right triangle.
improper fraction a fraction whose numerator is greater than or equal to its denominator. A fraction greater than or equal to 1.
integers positive or negative whole numbers and the number zero
interior angle an angle on the inner sides of two lines cut by a transversal
intersection the elements that two (or more) sets have in common
irrational numbers numbers that cannot be expressed as terminating or repeating decimals
isosceles triangle a triangle with two equal sides
least common denominator (LCD) the smallest number divisible by two or more denominators
least common multiple (LCM) the smallest of all the common multiples of two or more numbers
like terms two or more terms that contain the exact same variables

line a straight path that continues infinitely in two directions. The geometric notation for a line through points A and B is $\overleftrightarrow{AB}$.
line segment the part of a line between (and including) two points. The geometric notation for the line segment joining points A and B is $\overline{AB}$. The notation $\overline{AB}$ is used both to refer to the segment itself and to its length.
major arc an arc greater than or equal to 180°
matrix a rectangular array of numbers
mean the quantity found by adding all the numbers in a set and dividing the sum by the number of addends; also known as the *average*
median the middle number in a set of numbers arranged from least to greatest
midpoint the point at the exact middle of a line segment
minor arc an arc less than or equal to 180°
mode the number that occurs most frequently in a set of numbers
monomial a polynomial with one term, such as $5b^6$
multiple a number that can be obtained by multiplying a number x by a whole number
negative number a number less than zero
numerator the top number in a fraction. 3 is the numerator of $\frac{3}{7}$.
obtuse angle an angle that measures greater than 90°
obtuse triangle a triangle with an angle that measures greater than 90°
odd number a number that cannot be divided evenly by the number 2
order of operations the specific order to follow when calculating multiple operations: parentheses, exponents, multiply/divide, add/subtract
ordered pair a location of a point on the coordinate plane in the form of (x,y). The x represents the location of the point on the horizontal x-axis, and the y represents the location of the point on the vertical y-axis.
origin coordinate point (0,0): the point on a coordinate plane at which the x-axis and y-axis intersect
parallel lines two lines in a plane that do not intersect
parallelogram a quadrilateral with two pairs of parallel sides
percent a ratio that compares a number to 100. 45% is equal to $\frac{45}{100}$.
perfect square a whole number whose square root is also a whole number
perimeter the distance around a figure
permutation the arrangement of a group of items in a specific order
perpendicular lines lines that intersect to form right angles
polygon a closed figure with three or more sides
polynomial a monomial or the sum or difference of two or more monomials
positive number a number greater than zero
prime factorization the process of breaking down factors into prime numbers
prime number a number that has only 1 and itself as factors
probability the likelihood that a specific event will occur
product the result of multiplying two or more factors
proper fraction a fraction whose numerator is less than its denominator. A fraction less than 1.
proportion an equality of two ratios in the form $\frac{a}{b} = \frac{c}{d}$

Pythagorean theorem the formula $a^2 + b^2 = c^2$, where *a* and *b* represent the lengths of the *legs* and *c* represents the length of the *hypotenuse* of a right triangle

Pythagorean triple a set of three whole numbers that satisfies the Pythagorean theorem, $a^2 + b^2 = c^2$, such as 3:4:5 and 5:12:13

quadratic equation an equation in the form $ax^2 + bx + c = 0$, where *a*, *b*, and *c* are numbers and $a \neq 0$

quadratic trinomial an expression that contains an x^2 term as well as an *x* term

quadrilateral a four-sided polygon

quotient the result of dividing two or more numbers

radical the symbol used to signify a root operation; $\sqrt{}$

radicand the number inside of a radical

radius a line segment inside a circle with one point on the radius and the other point at the center on the circle. The radius is half the diameter. This term can also be used to refer to the length of such a line segment. The plural of *radius* is *radii.*

range all the solutions to $f(x)$ in a function

ratio a comparison of two quantities measured in the same units

rational numbers all numbers that can be written as fractions, terminating decimals, and repeating decimals

ray half of a line. A ray has one endpoint and continues infinitely in one direction. The geometric notation for a ray with endpoint *A* and passing through point *B* is $\overrightarrow{AB}$.

reciprocals two numbers whose product is 1. $\frac{5}{4}$ is the reciprocal of $\frac{4}{5}$.

rectangle a parallelogram with four right angles

regular polygon a polygon with all equal sides

rhombus a parallelogram with four equal sides

right angle an angle that measures exactly 90°

right triangle a triangle with an angle that measures exactly 90°

scalene triangle a triangle with no equal sides

sector a slice of a circle formed by two radii and an arc

set a collection of certain numbers

similar polygons two or more polygons with equal corresponding angles and corresponding sides in proportion.

simplify to combine like terms and reduce an equation to its most basic form

slope the steepness of a line, as determined by $\frac{\text{vertical change}}{\text{horizontal change}}$, or $\frac{y_2 - y_1}{x_2 - x_1}$, on a coordinate plane where (x_1,y_1) and (x_2,y_2) are two points on that line

solid a three-dimensional figure

square a parallelogram with four equal sides and four right angles

square of a number the product of a number and itself, such as 6^2, which is 6×6

square root one of two equal factors whose product is the square, such as $\sqrt{7}$

sum the result of adding one number to another

supplementary angles two angles whose sum is 180°

surface area the sum of the areas of the faces of a solid

tangent a line that touches a curve (such as a circle) at a single point without cutting across the curve. A tangent line that touches a circle at point *P* is perpendicular to the circle's radius drawn to point *P*.

transversal a line that intersects two or more lines

trinomial a polynomial with three unlike terms, such as $y^3 + 8z - 2$
union the combination of the elements of two or more sets
variable a letter that represents an unknown number
vertex a point at which two lines, rays, or line segments connect
vertical angles two opposite congruent angles formed by intersecting lines
volume the number of cubic units inside a three-dimensional figure
whole numbers the counting numbers: 0, 1, 2, 3, 4, 5, 6, . . .
zero-product rule if the product of two or more factors is 0, then at least one of the factors is 0.